ERNEST BRAUN

FROM NEED TO GREED

The Changing Role of Technology in Society

ÖSTERREICHISCHE AKADEMIE DER WISSENSCHAFTEN
MATHEMATISCH-NATURWISSENSCHAFTLICHE KLASSE

ERNEST BRAUN

From Need to Greed

The Changing Role of Technology in Society

Verlag der
Österreichischen Akademie
der Wissenschaften

Wien 2010

Vorgelegt von w. M. FRITZ PASCHKE in der Sitzung am 12. November 2009

Umschlagbild:
Manuela Kaitna

Umschlaggestaltung:
Manuela Kaitna

British Library Cataloguing in Publication data
A Catalogue record of this book is available from the British Library

Die verwendete Papiersorte ist aus chlorfrei gebleichtem Zellstoff hergestellt,
frei von säurebildenden Bestandteilen und alterungsbeständig.

ISBN 978-3-7001-6916-1

Gesamtherstellung: Ferdinand Berger & Söhne Ges.m.b.H., A-3580 Horn

Printed and bound in the EU

http://hw.oeaw.ac.at/6916-1
http://verlag.oeaw.ac.at

CONTENTS

INTRODUCTION

Technology is at the very centre of human activities. Historical periods are known by the leading technologies employed in them. Technology largely determines the way societies and individuals live. Our present period is often called post-industrial because production technologies are so advanced and so automated that vast quantities of goods are produced with little employment of labour. The age is also known as the age of information technologies (IT) because these modern technologies are dominant in shaping much of our present economic life.

Technology has made humans the undisputed dominant species on earth. Humans without technology are feeble creatures – with technology they conquer all before them. Yet our seemingly omnipotent technology is the greatest threat to the very existence of life on earth. We are destroying our natural habitat to an unprecedented and highly dangerous extent. With the aid of technology we have created enormous wealth, and yet we try hard to become even wealthier. We regard technology and technological innovation as the means toward the end of economic growth. Technological innovation is regarded as the main source of economic growth, and economic growth has become the very pinnacle of our desire. Economic growth means increased wealth and we tend, falsely, to identify increased wealth with increased happiness. Thus we are torn between praise and damnation of technology and charge our political system with fostering technology and yet curbing its apparent excesses. We are aware that we need to tread a narrow path between support and control, yet find it rather difficult to find the right compromise, to steer the right course between high hope and deep fear.

Despite the central role that technology plays in human affairs, there are very few books that attempt to clarify the way technology develops and the way it contributes to the development of society. These are the themes that have been at the centre of my interests for the past thirty years. The present book represents an attempt to shed light upon relationships between social and technological developments.

The method chosen is the historical approach. I believe that viewing the historical development of technology and of society in parallel is the most profitable way of gaining insights into the mutual relationships and influences.

Perhaps I am taking the historical approach to its extreme in beginning with the earliest dawn of humankind and ending with the present and a little beyond. It seems to me that this approach is likely to yield new insights and it is up to my readers to judge whether this is the case.

The book may be viewed in two alternative ways. It may be seen as a history of technology, albeit sketchy and selective, with remarks upon the social conditions and interactions between technology and society at any given historic period. Alternatively, and preferably, it may be viewed as an attempt to clarify the basic relationships between technology and society, illustrated by historic examples. I accept the dual nature of the book but hope that the attempt to clarify the interactions between technology and society will be seen to have the upper hand. These relationships are certainly uppermost in my mind.

My objective in writing this book is to achieve an overview of the ways in which technology shapes society and, in turn, is shaped by society. I aim to achieve some general understanding of the interaction and interdependence of technology with the society that creates and uses it. I hope that improved understanding of the mechanisms of interaction between society and its technology may lead to more beneficial and less destructive uses of technology.

Technology is at the core of all material and economic human activities and there can be no doubt that such activities dominate much of our thinking, all of our work, and much of our time.

Before turning to the development of technology and society through the ages, we should define our terms. I shall dodge the problem of defining society. Nearly all of us know what society is and very few subscribe to the view expressed by one recent politician, who claimed that there was no such thing as society. I shall leave each reader with his or her own understanding of what constitutes human society. To define this apparently simple concept unequivocally and comprehensively is fraught with too many difficulties to make the attempt worthwhile in the context of this book. In highly simplified and general terms, society is the community of humans living at a given time in a given geographic or organisational entity. Each society is characterised by certain prevailing relationships, legal frameworks, hierarchical and power structures and, of course, by the technologies it uses.

We next need to define technology, and this is a task that I do not feel free to dodge. The Encyclopædia Britannica (2000 edition) defines technology as "*the means or activity by which man seeks to change or manipulate his environment*". The word technology is thus a noun that describes certain classes of artefacts, as well as certain classes of activities. An alternative definition is "the application of scientific knowledge to the practical aims of human life". I regard this definition as unsuitable because it applies only to high technology and not to older, simpler, technologies. There is some justification for such a definition in certain languages, e.g. German, because the word 'Technik' can be used to describe simple technologies and the word 'Technologie' to describe science-based technologies. In my own definition *technology is the production and use of artefacts for human purposes.* Alternatively we can say that *technology is the way and the means of achieving practical human purposes.* The essential feature of technology is the production and use of artefacts. The artefacts involved are tools or machines or chemical entities. The term 'practical' (we might say 'material') distinguishes the purposes achieved by technology from purely spiritual, educational, cultural, or organisational purposes.

Some higher animals use objects found in nature, such as twigs, stones or leaves to achieve practical purposes like fighting, crushing nuts or soaking up inaccessible water. We might term these objects tools, but they are not purpose-made artefacts. They are natural objects, maybe slightly modified, but not planned, designed and produced for a specific task. Hence my emphasis on the use of artefacts: only humans manufacture artefacts for the achievement of their purposes. The distinction between producing and using a chipped stone as a tool, as early humans did, and using chewed leaves to sponge up water, as chimpanzees do, is rather slender, but perhaps the difference between early humans and apes was rather slender.

Definitions serve the purpose of describing certain objects or activities to the exclusion of all others. Too broad a definition serves no useful purpose. I disagree with S. Macdonald[1] when he says "…it is quite possible to view change in administrative or managerial practices as technological change …." Of course it is possible, but it is not helpful. We do not wish to define technology so broadly as to make any bishop into a technology manager, although his activities are in part managerial and do serve some human purpose. On the other hand, there can be no doubt that a bishop, just as any other member of modern human society, uses technology in many of his daily tasks. So ubiquitous has technology become that we use it all the time in almost all that we do. The simplest definition of technology as "*the way humans achieve their material purposes*" is not far off the mark.

Cooperation between individuals and groups has been a vital feature of human societies since the dawn of humankind. Even very early hunters, to give an example, cooperated in hunting animals and used various weapons, such as wooden spears, in a concerted effort to trace, chase, and kill. The use of the spear is technology, whereas the organisation of the hunt, and the various ruses used, we shall term *social organisation*, rather than technology. The activity of hunting, as so many social activities, requires technology as well as social organisation. Indeed the use of technology is generally associated with, or even dependent upon, social organisation.

The word technology has a somewhat high-powered ring to it. We often use the word *technique* instead, particularly to describe details of technical procedures. For the Stone Age, it is best to limit the use of the

1 S. Macdonald, p. 27, in S. Macdonald et al., 1983

word technology to the general methods of manufacture of tools and other artefacts, while using the word technique to describe the detailed ways of using tools. We may speak of the techniques of butchering carcasses, or of providing shelter, producing clothing, and so forth. We also use the word technique to describe details of any particular method of, for example, knapping. The basic technology is using a hammer to shape a stone tool out of a natural blank, perhaps a large pebble of flint. The detailed way of splitting the stone, and of chipping flakes off it to obtain the desired tool may be termed technique. Other derived terms are the adjectives technical or technological and the nouns technician or technologist and several more.

Both words, technology and technique, are nouns. To describe actions associated with technology, as for all actions, we need verbs. Some verbs describing technical actions are directly associated with the word describing the tool used. Thus hammer, and saw are nouns and verbs; the verbs scrape and bore are associated with the nouns scraper and borer. But we sew when using needle and thread, we thrust or throw when using a spear and we build or construct a shelter when using a variety of tools and techniques.

Much linguistic development is associated with the development of technology. A very large number of our words are used to describe artefacts and technology, and the number of words increases with technological progress. Computer and information technology have contributed a large number of English words that have come into use throughout the world. Examples are computer and computing, CD (compact disc), DVD (digital video disc), Internet, e-mail, LAN (local area network), modem, and many more. Some older words have acquired additional meanings: surfing, virus, café, or web.

Tools are means to an end. Some tools are used to make other tools, and thus satisfy our needs less directly. Even in earliest times humans used stones as hammers or anvils to make stone tools that were used to produce food or achieve some other desirable objectives. We may regard weapons as a special form of tool used either for hunting – and thus the production of food – or to kill enemies in the pursuit of some economic or social objective. In all cases tools are used to enhance, amplify or supplement our innate abilities. A hammer adds hardness to our hands and helps to focus muscle-power; an aeroplane enables us to fly; a computer amplifies our memory and adds speed to our mathematical powers.

The prime purpose of all living creatures is survival, and humans are no exception in this respect. Thus the needs that arise directly out of the primary purpose of survival are primary needs and were the first to be served by the earliest technology. Gradually technology became able to serve further needs and its focus shifted from aiding sheer survival to becoming an essential economic force. In modern societies technologies mainly serve to make money, to help the economy to grow. The focus of technology has shifted from satisfying primary needs to the satisfaction of human greed. This is true for producers of technology who are after profits and consumers of technology to expect technology to satisfy ever more sophisticated and far-ranging demands. How this happened is the story of this book.

Over the years, I have derived much benefit and pleasure from numerous discussions with students and colleagues on the subject matter of this book. All of them, too numerous to name, I thank wholeheartedly. My thanks are also owing to my wife for putting up with my frequent physical and mental absences and my obsessive preoccupation with the task of writing this book. Special thanks to Geoffrey Boyfield and Helge Torgersen for reading an earlier version of the whole manuscript and offering much helpful comment.

Hadleigh, March 2009 Ernest Braun

Chapter I

In the Beginning

We may place the beginning of technology with the first human-like creatures making simple stone tools. Homo erectus, who emerged in Africa some 1.8 million years ago, is considered as the earliest toolmaker[1]. Modern humans, known as Homo sapiens, also originated in Africa some 100,000 or more years ago and came to Europe roughly 40,000 years ago. The first hominid in Europe was Homo heidelbergensis, from about 700,000 years ago, followed by Homo neanderthalis, who disappeared some 30,000 years ago. Hominids that pre-dated the first toolmakers undoubtedly used natural objects as tools, but Homo erectus was the first to make tools. Many primates and other animals use natural objects as tools, for example sticks as weapons or stones for cracking nuts; and even occasionally modify such objects, by chewing or other manipulations, to make them more suitable for the task in hand. The modifications, however, are never very substantial and tools are not made for future use, but only for an immediate purpose. Man, on the other hand, perceives of possibilities of using natural raw materials to make objects that are very different from the original materials, though even in the late Stone Age some natural objects were used as tools, such as antlers for pickaxes and bovine shoulder blades for shovels, though not without a considerable amount of cleaning and preparation. Only humans appear to have the imagination necessary to plan major modifications of natural objects for particular practical purposes. Only humans appear to have the foresight to make tools for future use, for occasions that might arise at some future time. Foresight and imagination are part of the tool-making process. Humans have two further abilities that, in contrast to animals, enable them to make intricate artefacts. One is a superior ability of coordination between visual and tactile sensations or, in other words, superior coordination between eyes and hands. Human dexterity, controlled by vision, far surpasses that of any other creature. The second ability is concentration. The concentration span of humans exceeds that of animals and thus humans are able to produce artefacts that require a great deal of effort, sustained over long periods. Remarkable as some feats of animal construction are - we may think of the elaborate dams built by beavers or of birds' nests – they are no match for the human facility of producing artefacts.

Tools are means to an end. Indeed technology may be defined as *the material ways and means to achieve some human purpose*. The prime purpose of all living creatures is survival, and humans are no exception in this respect. Thus the needs that arise directly out of the primary purpose of survival are primary needs and consist of:

a. the need for food and water, may include hunting;
b. the need for protection from cold, be it by clothing or by shelter. The latter may consist of natural caves or of man-made objects;
c. the need for protection from predators and, possibly, fellow humans.

The earliest technologies addressed the primary needs and were an essential aid to the survival and development of the human race.

[1] Homo habilis is a slightly older form of hominid that made tools, but is generally regarded as less important. The pre-history of humans is very complex, ill-documented by fossil finds, and somewhat controversial. The matter is not central to our arguments and will not be discussed in detail.

Throughout the Lower Palaeolithic period, from about 2.5 million years ago to roughly 200,000 years ago, only these categories of purpose played any role. In the Upper Palaeolithic, Mesolithic and Neolithic periods, covering from 200,000 years ago to something like 3,000 BC, further purposes gradually gained in significance. Present day technology, although incomparably more extensive, complex and sophisticated, still fulfils our basic needs, though with much greater complexity and variety than ever before. And, of course, modern technology serves many more categories of purpose and enables vastly increased numbers of humans to survive.

Palaeolithic societies consisted of small groups of hunters and gatherers who lived a nomadic life and used tools merely as aids to survival. Technology was born out of the need for survival. Hominids - and humans - are physically ill equipped to survive unaided in hostile environments. They are neither fast, nor powerful, nor endowed with effective natural weapons. Their only equipment for survival, or competitive advantage, is their intelligence and their ability to make and use tools. Tools to help them gather food, scavenge and hunt for meat, build shelter and tame fire, and to produce clothing. Only with the help of these early technological achievements could they survive and thrive against the odds of being eaten by predators and of starving or freezing to death. Technology assisted hominids in their struggle for survival since the days when they first made their appearance on earth. Indeed this aspect of technology, to assist humans in their struggle for mere survival, is a vital part of technology to this day, even though this task has been overshadowed by many other functions of contemporary technology. Need was the mother of early technology and imaginative dexterity was its father.

In the beginning was the use of natural objects as tools; the manufacture of tools was a crucial next step. It is this step that humans alone have taken. Many animals use, and sometimes slightly modify, natural objects as tools – man alone makes tools. Man is a toolmaker or, to put it the other way, tool-making makes man. It is easy to imagine that some of the more intelligent of these early hominids would first look for sharp stones in nature that could scrape, dig and cut more effectively than blunt stones; they would then hit upon the idea that sharp stones can be obtained by hitting blunt ones with another stone. Gradually they would perfect the manner of producing such sharp stone flakes that constitute early stone tools. They would learn how to select the best raw materials, such as flint, and would learn how best to hit it to produce sharp flakes. For many millennia the development of technology was very slow. It was not till Neolithic (new Stone Age) times, about 10,000 years ago till about 3,000 BC that technological development accelerated, the techniques of producing stone tools were perfected and the variety and quality of the tools improved greatly.

Palaeolithic man was a nomad who moved in cycles determined by the seasons. He had to follow the migrations of game herds and the ripening of edible plants in various locations. He used caves for winter shelter and simple temporary shelters during his migrations. It is known, for example, that many sites in the British Isles, where human remains and other traces of human presence were found, were not continuously occupied. If no caves were available, he was forced to build more permanent shelters from what materials he could find. Occupation of different territories depended on the state of glaciation. Humans were nomads in two senses – the usual sense of having no permanent abode but wandering from place to place in search of food; and also in the sense that climatic conditions limited the regions in which humans could survive and forced them to move with the ebb and flow of the ice. We might speak of short period and long period wandering. During periods of greatest glaciation many northern parts of Europe became uninhabitable, while milder interglacial periods allowed habitation of more northerly regions. The fauna and flora, and even the landscapes, changed dramatically with climatic changes and all that contributed to enabling, or not enabling, humans to occupy a region. Climatic conditions were a major determinant of the way humans lived and what food was available to them. Their spread through the world was similarly dictated by geographic and climatic conditions. It is thought, for example, though disputed by some, that humans reached America when that continent was practically connected to Siberia. Similarly, humans first inhabited Britain during a period when it was still firmly part of continental Europe.

We have said that humans survived in hostile environments by using their intelligence and their hands, equipped with natural tools at first and artefacts later. It was only with the use of artefacts, however, that humans managed to survive in large numbers and, eventually, to spread across the world. Tools allowed them to dig for edible roots, to scavenge efficiently for meat, to use sticks, branches, stones, grass and hides to produce temporary shelters. Intelligent cooperation enabled them to outwit their prey and hunt for large animals and, in the fullness of time, to kill not only weak animals but also those in their prime. Early on the ability to produce tools and to plan ahead and to cooperate in common endeavour were the most important uses of human intelligence. Planning enabled hominids to arrange their days so as to reach suitable shelter when needed, to plan the hunting of small and, later, larger animals, to obtain water at times and in places where they avoided excessive danger from predators. At some time they learned to make arrangements to keep their fires going. It is likely that very early on they depended on natural ignition by lightning, though later they learned not only to tend, but also to light fires. This had two effects: they learned to cook food and this, over many generations, caused their digestive apparatus to shrink and their brains to grow. Cooking replaces part of the digestive process and thus, indirectly, advances intelligence.

The second effect was that fire enabled humans to survive in colder climates. It is likely that the mastery of fire was a pre-condition for human migration into most parts of Europe. The Cro-Magnons certainly spent their winters in caves warmed by fires during the coldest period of the last ice age. They could venture out only in properly tailored warm clothing made of animal skins.

Nobody knows exactly when language developed, but undoubtedly some form of communication was part of pre-human society from very early on. After all, apes communicate, even if they have nothing that we recognize as language. Gestures, shrieks and grunts and facial expressions must have become ever more complex and over thousands of years the first simple languages emerged. No doubt anatomical development of the larynx and pharynx was a pre-condition for articulate speech. It is thought that Homo erectus developed the anatomical potential for speech roughly 300,000 years ago.

We must stress the great uncertainty involved in all our knowledge of pre-history. The only certainty is our uncertainty. Stories and theories are conjectured on the basis of very few finds and experts disagree over many interpretations. Though lithic artefacts can be identified with some confidence, methods of dating very ancient sites and finds are inaccurate and often contradictory. As far as this book is concerned, I am trying to tell a more or less coherent story based on as much consensus as I can find, but it is not my ambition, or within my competence, to provide an authoritative detailed account of the development of the human race. My ambition is to tell a consistent tale of technological developments and correlate these with reasonably well-founded facts on human society.

The road from producing crude stone flakes to the production of sophisticated tools made of stone, wood, horn, bone or antlers is a very long one and is intimately connected with the development from Homo erectus to Homo sapiens. One of the many problems in studying archaeological artefacts is that most of them have not survived to the present. In fact almost anything made of hide, wood, bone or antlers and other biological materials has long since disappeared. What remains are stone tools and even they pose problems. It is difficult to distinguish with certainty between natural stone flakes and artefacts and also difficult to know what the ravages of moving ice sheets, volcanic eruptions, flowing rivers and other forces of nature have done to the shapes and locations of lithic artefacts.

We must use our imagination to conjecture what uses the earliest tools were put to. Whereas small animals could be killed with sticks, and nuts or bones could be crushed with pebbles, a sharp instrument was needed to separate the skin from the flesh and the meat from the larger bones and to cut the meat into manageable portions. Thus scrapers and knives became an essential item in the tool-kit of early humans. Some form of axe was needed for chopping down trees or shrubs, whether for the production of weapons, or for firewood, or for the construction of simple shelters. The hand-axe, a ubiquitous tool of a slightly later period, represents a combination of scraper, knife and axe. For killing larger animals, and possibly fellow humans or predators, a spear was needed. Early spears were made of wood, with the tip sharpened with an axe; later stone tips were fixed to a wooden shaft by various methods.

During this early period of technological development there was neither specialisation nor standardisation. Everybody made their own tools as best they could and each tool was slightly different from all others. More standardised and more specialised tools became prevalent later, when small bands of humans merged into something approaching society and the knowledge of making and using tools became social knowledge, shared by members of a given social group.

The earliest stone artefacts consisted of naturally sharp-edged flakes broken off a larger stone, either by hitting a pebble with another stone or by hitting it on a stationary larger stone or rock, an anvil. We call this very earliest stone technology Oldowan technology because the first examples were found in the Oldovai Gorge in East Africa. These early tools might have helped with gathering wild crops, but their main utility was in scraping skin and bones off meat and perhaps cutting meat, hides, and wood. Meat was probably obtained mainly by scavenging, though the tools, coupled with the use of sticks, may have helped to kill and butcher small animals.

Homo erectus spread from Africa into Eurasia. It is estimated that the first such humans reached Europe about half a million years ago by way of North Africa and the Middle East. The stone technology they left behind is known as Acheulean technology, named after St. Acheul, near the Somme river in Northern France, where the first archaeological finds were made. The principal tool of the period is the hand-axe. It has been found in many regions of the world and comes in many sizes and slightly different shapes. Essentially it is made from a natural piece of stone and first hammered roughly into shape, then shaped more accurately with a lighter hammer, perhaps made of bone. Hand axes were the principal universal tools over a very long period, supplemented mostly by sharp stone flakes used for cutting and scraping meat and skins.

The Acheulean technique was later refined into the so-called Levallois technique. In this method a stone core was carefully trimmed in such a way that a final well-placed and angled blow would detach a large flake directly usable as an implement. The core was discarded. Such tools are known as unifacial tools because one surface was flat. They could be large and wedge-shaped with a straight edge, to be used as choppers or cutters, or they could be pointed for use as burins or borers. Tools with serrated edges, an early form of saw, have also been found. It is thought that these tools were used to cut and shape wood, but also to prepare skins for clothing or temporary shelter. The Neanderthals used assemblages consisting of hand axes and various tools made by the Levallois technique and their technology is collectively known as Mousterian. If we regard hunting as a technology, it certainly developed during this period. Neanderthals learned to hunt large game such as reindeer, mammoth, bison and wild horses. This shows that the Neanderthals lived in somewhat larger groups and were able to cooperate in hunting and other activities. Presumably they were able to communicate by speech. Neanderthals became adept at survival in cold regions, which means that they must have mastered the art of keeping, and probably lighting, fires.

We should not be surprised that it took so many millennia to develop the Levallois technique. We must remember that human intelligence and innate skills, especially eye-hand coordination, needed to develop and that this is a very slow biological process. Another important factor is the very small size of groups that formed human society and the very limited interaction between groups. As most learning is social learning – we learn from each other and develop or discover very little for ourselves – limited social intercourse puts a considerable brake on technological development. Learning requires interaction and communication between many individuals and if such interaction is rare and limited, learning proceeds but slowly.

Tools found in different locations, dating from the same period, are similar, but somewhat different. This fact can be used to argue either the case of a simple technological determinism or the case of social learning. It could be argued that the logic and possibilities of using stones for practical tasks determines, to a considerable extent, what shapes shall be prepared. We shall return to this discussion later, when more advanced technology and more coherent societies form a better basis for discussion. At this early stage it is clear that the technological options were pretty limited, as was social intercourse. Nevertheless, the fact that regional differences are discernible tends to show that some social learning was involved even in these early stages, although the technological determinism was probably stronger than the social forces.

Early Palaeolithic society had a very limited range of social interactions, let alone institutions. These early humans probably lived in bands of gradually increasing size, though below about 100 members. They had no known rituals and interaction between bands was limited to occasional chance meetings. Even these limited occasions were probably enough for some exchange of technical know-how and, thus, to social learning. Within a band, social learning probably played an important role as each member could observe all other members in their daily tasks and the naturally more talented stone-workers passed some of their skills to younger members of the group. Whenever a group uses a technology, this passes into the ownership of the group and becomes social knowledge. Shared knowledge is social knowledge. Throughout the Palaeolithic period social life increased and by the time the Cro-Magnons took over, groups had increased in size, had developed some forms of social institutions and had more intense contact with more distant populations.

The Cro-Magnons, who succeeded the Neanderthals as the main Europeans, made substantial contributions to technological progress. One of the great advances made by them was hafting. Attaching a stone tool to a wooden shaft would increase the available kinetic energy considerably and thus enable the user to cut and shape much larger trees and timbers. Attaching smaller tools to some form of handle or holder would render their use much more convenient. Many new varieties of tools were introduced to serve specialised purposes. The Cro-Magnons produced spear-points from stone or bone and were able to attach these to wooden shafts. The surface finish of the tools improved, as they were retouched with light hammer blows and by grinding and polishing. The Cro-Magnons developed burins, which enabled them to produce eyed needles from slivers of bone or antler. This enabled them to produce clothing from animal skins and furs. Remembering that the Cro-Magnons survived the coldest periods of the last ice age, their clothing must have been very well made.

It is possible to scavenge for meat, but in order to obtain undamaged skins for the production of clothing, it is necessary to kill animals. They could either be driven toward some precipice and the carcases collected at the bottom, or they could be trapped in deep holes adequately disguised, or they could be killed with lances, spears and axes. Both for hunting and for the preparation of meat and of clothing suitable tools had to be developed. But tools were not enough; to trap or hunt large animals cooperation between several hunters is necessary. Thus a pre-condition for the production of effective clothing was the development of effective social organisation. If you have only very simple tools, you have to cooperate with others in order to kill large animals. Fishing also provided an important source of protein. Fishing was carried out either with the aid of rouses, or nets made from natural fibres, or fishhooks that were made mostly of bone.

The Cro-Magnons were biologically modern humans and it is thought that they may be the remote ancestors of some humans living today. They flourished in the Upper Palaeolithic, from about 35,000 years ago to about 10,000 years ago. Both Neanderthals and Cro-Magnons buried their dead, though it is impossible to know whether the reasons for this were ritual or hygienic. The Cro-Magnons almost certainly developed some form of faith and of ritual.

Cro-Magnons were the first humans to be known to have produced art. They produced cave paintings, depicting mostly hunting scenes and found mainly in France and Spain. The cave paintings provide many clues to the life of these societies. Hunting scenes are an important feature and show some of the weapons and techniques used, as well as the animal species hunted. We do not know for certain whether these pictures were associated with rituals, perhaps incantations to the spirits or gods to provide good hunting, or whether they were purely the result of artistic impulses. In any case, they provide proof of considerable facilities for abstract thought. The Cro-Magnons obviously had aesthetic feelings, shown in decorations engraved on tools and weapons. They also carved small sculptures, mainly of big-breasted pregnant women. One assumes that these were used in some form of fertility rites.

Perhaps the most famous, though not the oldest, of these cave paintings are those of Lascaux, in the Dordogne region of France. The vast cave contains a large array of wonderful paintings, mainly of animals such as aurochs (wild cattle), horse and deer. They are dated to about 15,000 BC. It is obvious that by this time humans in this region had not only the capacity for abstract thought and a desire for artistic expression, but also had a set of techniques that enabled them to produce these spectacular images. They needed pigments, pestles and mortars, brushes, lamps and scaffolding, quite apart from great skill and sufficient time, presumably available during the long cold winter. It is possible that some of the paintings served a ritual designed to influence the outcome of hunts. Apart from the realistic images, the paintings also contain a number of abstract geometric patterns, whose significance is not understood. Was it an early script or early abstract art?

The cave paintings are an early instance showing technology serving a class of purpose that falls outside primary needs. Such classes of purpose include art, ornamentation, ritual, play and religion. From their beginning with cave paintings, engraved ornaments and simple sculptures, this class of cultural needs served by technology expanded hugely from the Stone Age to our modern days. One might conclude that as the daily struggle for survival became more successful and no longer occupied all the time and energy available to humans, their minds turned to other things, to a different direction of their creative abilities. Homo faber is not only a maker of tools, but also a maker of artistic artefacts. We do not know whether these activities originate from a desire to enlist the help of spirits in practical endeavours, or whether they were created just for the pleasure of creation and contemplation.

The Upper Palaeolithic, with Homo sapiens the dominant and probably the sole human species, saw the introduction of further new improved technologies. Stone tools were now made in two stages. The first stage consisted of producing long slender blanks, or blades, that could then, in a second stage, be worked into all kinds of specialized tools. The long flint blades were of trapezoidal cross-section. This method of producing stone tools was the culmination of the development of lithic technology and is known as the blade-tool industry. It could produce long slender knives or small tools, known as microliths. Both knives and microliths could be attached to handles made of bone, wood or antler, using resin as glue. It is probable that by this time some degree of specialisation had developed and it is possible that both blades and finished tools were manufactured by specialists and traded.

The Neolitic age took developments in lithic technology even further. One of the hallmarks of Neolithic stone tools is the superior quality of their surface finish. They were well finished by light hammer strokes and then ground and polished, either on a rock or with abrasive powder, so that their appearance was smooth and their cutting edges sharp and even. The smooth finish not only made these tools into more desirable objects, it also increased their efficacy and ease of use by sharper edges and reduced friction. A further hallmark of Neolithic tools was the use of superior materials. Whereas previously the materials used came mostly from the immediate surroundings, in Neolithic times superior flint was traded over quite long distances and much of it came from flint mines. This required the development of mining techniques and the transportation of the mined products to different places for knapping and use. A further excellent and rather rare raw material for tools, obsidian (a volcanic glass) appears to have been obtained from quite faraway places, either by exploration, by exchange with travellers, or by trade.

A major invention of this period was the bow. The bow may be looked upon as a device that stores and accumulates the energy of the archer's muscles and, on release, imparts greater momentum to the arrow than the archer could impart unaided. The bow and arrow were destined to become one of the principal weapons for a very long period, probably more than 10,000 years, until the 15th century AD. The bow and arrow proved as useful in hunting as in combat. It spread rapidly to all populated regions, showing that by then contacts between different populations were well established, thus enabling a superior technology to diffuse quite rapidly.

Throughout the millennia that we have described, albeit briefly, natural selection worked in favour of higher intelligence and of greater prowess in making and using tools and weapons. More intelligent individuals stand a better chance of living longer, and of producing more offspring, than their less intelligent contemporaries. This is the classical evolutionary pressure toward the development of higher average intelligence of a population. The less able have lower chances of survival and of procreation than the more able and thus the species evolves toward greater ability. In particular, it is technical prowess, superior tool making, which improves the chances of survival, though undoubtedly physical prowess was of great importance for survival under harsh conditions. Growing average intelligence causes greater average chances of survival and, thus, leads to an increase in population. Improved tools, which help to provide more food, lead to the demand for ever more food and thus to a demand for ever better tools. A feedback loop between better technology, growing populations and the need for ever improved tools evolves. Man the advanced toolmaker could survive in places and in numbers that were inaccessible to humans without good quality artefacts.

Though the technology reached by the end of the Palaeolithic, or even the Neolithic, was still very simple when viewed with modern eyes, the development of technology up to this period already demonstrates three principles of the development of technology that have remained valid to this day.

1. All technologies and technological products improve with further development. They become more effective and efficient, more comfortable to use, more streamlined. We can define one or more figures of performance for technologies, and these figures increase with the development of the technology. The improvement continues until the particular technology becomes obsolete, either because a superior type of technology re-

places it, or because the need it had served disappears. Stone technologies largely disappeared because they were replaced by superior metal technologies. An interesting figure of performance for lithic technology is the length of cutting edge that can be obtained per unit weight of raw flint or similar material. The increase throughout the life of the technology is quite spectacular (Table 1.1)

Length of cutting edge produced from 500g of flint or similar material	Years ago	Human Species	Type of Technology
80 mm	2,000,000	Homo habilis	Oldowan
300 mm	300,000	Homo erectus	Acheulean
800 mm	100,000	Neanderthal	Mousterian
9,000 mm	10,000	Homo sapiens	Blade

Table 1.1 Figure of performance of lithic technology (Source: B. M. Fagan, 1998, p. 87)

2. Technology becomes able to satisfy ever more human needs, or serve ever more human purposes. At first it only helped with the provision of food gathered or scavenged. Later it helped with hunting and with the better utilisation of food by making cooking possible. It helped with the provision of shelter, clothing, warmth and light. Technology provided weapons not only for hunting, but also for defence against predators and, soon enough, for inter-human warfare. Before long, at least as early as 16,000 years ago, technology became instrumental in producing symbolic, ritual and decorative objects. This shows clearly that needs expanded as possibilities to satisfy them became available. At first technology was driven only by elementary biological needs. Later, further needs were added and this addition of needs has expanded and accelerated to the present day.

The need for symbolism and ritual arose because the small scattered bands of humans had grown into an early form of society. Technology helped more people to live longer and thus their numbers increased and the small hordes became larger communities. Ritual is an expression of cohesive forces in society and technology made its contributions to ritual when it became able to do so.

3. As the purposes served by technology become more varied, so technology becomes specialised. Whereas the original stone flake served as a universal tool and the Acheulean hand axe served most purposes, later tools were made for specific uses. Scrapers, cutters, choppers, saws, diggers, needles, borers, spears, fishing hooks and nets, boats, huts, fireplaces, lamps, and so forth.

By Upper Palaeolithic times, the categories of purpose served by technology had increased to at least four, albeit all served at a rather elementary level, not much beyond mere survival. In addition to the previously listed purposes, technology now also provided artefacts in support of decorative, symbolic and ritual purposes.

The new Stone Age, the Neolithic age, began and ended at different times in different regions. Generally it was preceded by the Mesolithic, a transition from the Palaeolithic to the Neolithic, and succeeded by the metal ages – copper, bronze, iron. To give a rough idea, we can place the Neolithic between, say, 8,000 BC and 3,000 BC, though this may be out by a couple of thousand years in some places. By the beginning of the Neolithic, Homo sapiens had spread to most parts of the world, including Southeast Asia, New Guinea, Australia, Siberia and America. Most of the places could be reached either by land or by quite short sea routes. The intriguing question is what drove humans to roam all over the world. I suppose the main factor was competitive search for food and changing climatic conditions, but curiosity and the challenge of the unknown may have played a role.

The Neolithic brought about the most momentous change of all. The two major innovations defining the age were pottery and, above all, agriculture. Man learned to grow crops and to domesticate animals. This changed his world from one in which he had to roam to find food, into one in which he could settle into permanent settlements and produce food. The process was very gradual, with nomadic lifestyles co-existing with sedentary ones for a long period. The development of agriculture changed the economy from one in which everything that each human produced was consumed more or less instantly, into one where surpluses could be produced, thus enabling human society to create an infrastructure of specialised workers, including artisans, administrators, priests, and rulers. As long as each person instantly consumes the food that he/she finds, it is impossible to sustain people who are fully engaged on pursuits other than gathering or hunting for food. Specialists who do not produce food have to be fed by the surpluses of food producers.

As an aside, we may indulge in an interesting, though gruesome, speculation. If people who consumed all the food they could produce made a prisoner in any kind of war, the only sensible thing to do with the prisoner was to kill and, preferably, eat him or her. If, on the other hand, prisoners could produce more food than they needed to consume, then it was worthwhile to keep them as slaves. Thus slavery was born, and with it some of the darkest chapters in human history.[2] It is not known when the first wars were fought, or when hunting weapons were first used against fellow humans. It is likely that Cro-Magnons and Neanderthals fought over resources and equally likely that different later tribes of Homo sapiens, alias Homo bellicosus, fought each other for reasons of greed, lust for power, envy or some such.

How agriculture initially developed can only be surmised. The most likely explanation is that people observed that the seeds they gathered from wild oats or wheat and other grains could be sown and thus grown in greater densities for easier gathering. They probably also observed that some seeds can be harvested more easily than others, because they stay on the stalk for a while when already fairly ripe. Undoubtedly they also found that some seeds are larger than others and thus provide better food value. Such observations may have caused them to start selecting seeds for sowing and a long chain of development of cultivated crops may thus have started. Perhaps the first seeds were sown by accident, when some dropped on the ground and were observed to sprout. Perhaps people observed how plants seed themselves in nature and hit upon the idea to emulate this process. After all, they collected seeds and must have accumulated quite a lot of knowledge about them.

Apart from wheat and barley, a variety of other edible and useful plants were of importance in Neolithic times; with some of them growing wild, others coming under cultivation. Green vegetables, such as cabbage, lettuce, nettles, cress, beans and peas were planted, albeit in forms much closer to their natural forms than to the vegetables we see on modern market stalls. Carrots and radishes were among the root vegetables; and beans, peas and lentils among the pulses. Apples, pears, plums and cherries were important. Vegetable oils were obtained from walnuts, linseed, poppy-seed, rape and olives. Rye, oats, millet and rice were gradually added to the earliest cultivated cereals. A variety of plants served as raw materials for the production of artefacts: timber, palm trunks and reeds for building; gourds for vessels; flax, hemp and cotton for fibre. Boats and paddles were constructed from much the same materials.

The knowledge gathered by individuals soon began to circulate and became enriched in the process. Different people exchanged their knowledge and experience and thus spread, enhanced and refined it. Communal knowledge grew out of a large number of individual observations. By the time of the Neolithic, language was quite well developed and exchanges between people within and between communities had become widespread. Knowledge about planting deliberately for crops gradually accumulated and, at the same time, knowledge about cultivation must have advanced. Special implements, such as a variety of digging sticks and hoes, were developed as predecessors to the plough. Sickles with stone blades and wooden handles were developed and used for harvesting.

Eventually, knowledge about crop rotation and about fertilisation and the use of grazing animals became part of agriculture. The latter presupposed animal husbandry that developed at about the same time as planting and cropping methods. Again, we can only speculate how animals were first domesticated. Perhaps people hit upon the idea that herd animals could be driven into pens and thus made more accessible for meat, for wool and for milk. Wool was gathered by plucking and it is obviously much easier to do this with penned or tethered animals than with freely roaming ones. It is also much easier to preserve meat on the hoof, rather than preserve surpluses accumulated in a successful hunt. The production of milk and milk products was realised later as an added bonus.

---

[2] V. Gordon Childe, Early Forms of Society, pp. 43–44 in Singer et al., 1954

We may ask why agriculture developed at all. Was it simply a result of newly recognised opportunities, of factual observations, of inventions based on the study of nature? Or was there some pressure, perhaps of increasing populations, an increasing number of mouths to fill, which forced people to seek ways and means of increasing food production? Or did humans attempt to free themselves from the vagaries of weather and climate, to obtain not only more abundant, but also more secure and regular food supplies? I think the answer must be a bit of each. A constellation of circumstances in which a problem – increasing population and insecure supplies of food – and a possible solution – intensification of food production – came together. A need and a possibility to satisfy it are the fundamental ingredients of technological innovation. We cannot say which was the egg and which the chicken, all we know is that agriculture increased food supplies and this enabled the population to grow, but whether agriculture or growth came first we cannot say.

The fundamental reason for wishing to grow crops and domesticate animals is the need to obtain larger and, of equal importance, more stable and secure food supplies; to gain some independence from the whims of nature. The reasons why agriculture developed much earlier in some regions than in others are hotly disputed. The reasons may be either greater need or better opportunities in some regions compared to others, or a combination of both. It may be unusual pressure of population, or particularly difficult ecological and climatic conditions that render traditional food supplies inadequate or unstable. On the other hand, it may be that natural conditions were particularly favourable to the development of agriculture in particular regions.

It is generally agreed that the process of domestication of farm animals – initially goats and sheep – started in southwestern Asia in roughly 8,000 or 8,500 BC. The process of 'domestication' of wild plants, especially of wheat and barley, is thought to have started in the same region at about the same time. The valleys of the rivers Jordan and Euphrates were among the first regions of grain cultivation. In a sparsely populated world, with extremely slow and erratic communications, technological revolutions were spread out over a long period. It took about 8,000 years from the beginnings of farming to a virtually complete conversion of the world's population from hunting and gathering to farming. The dog forms a separate case and was domesticated long before the rise of agriculture. We must assume that the dog proved its utility by helping hunters and by guarding households. No doubt the bond of friendship between humans and dogs was forged in very early times.

The consequences of these developments are obvious and far-reaching. People who have fields and animals to tend lose their mobility. They have to settle near their fields and can build more permanent shelter. Thus the first permanent settlements with proper huts became the forerunners of recognisable villages with various communal arrangements. Apart from leading to a development of building techniques, using timber, earth, grass and hides, there were other far-reaching consequences. Or were they consequences, rather than parts of a jigsaw that formed a new picture? May we speak of causes and effects, or must we seek to describe constellations of circumstances that gave rise to a different society with different technologies?

We can be certain, though, that increased food production and the settled way of life led to entirely new and far-reaching social developments. It became possible to accumulate surpluses of food and either gather these as wealth or exchange them for other forms of wealth. As soon as wealth was created, it became possible for some families to accumulate more of it than other families. Even more important, land acquired a value and could be appropriated, either by communities or by individuals. Nomads roam and do not lay individual claims to land. During nomadic times, land was communal and did not have any wealth or ownership implications associated with it. It was the advent of agriculture that created the concept of land ownership. Once ownership claims were made, the foundation was laid for hierarchies. Ownership and hierarchies lead to disputes and to the need to settle them, thus some rudimentary form of a legal system had to be established. Wealth, hierarchies, and laws in their turn are the foundations of power and of political systems. In further consequence, power and wealth become incentives for conquest and war, thus creating a new constellation in which weapons become the technological aspects of military-political systems. However, the true development of hierarchies and all their consequences did not happen till the metal ages, though the Neolithic period was one of accelerating social and technological change.

It may be appropriate to formulate another fundamental law of technological development at this point, albeit in a preliminary form to be refined later. Whenever social change occurs, whether primarily because of changes in technology or for other reasons, new needs arise for technology to satisfy. A sedentary agricultural society requires new technologies, such as agricultural implements and permanent shelter.[3] Technological and social changes go hand in hand and fundamental social changes, however caused, require the implementation of technological change. We can turn this round and say that fundamental technological change needs to be accompanied by social changes.

One of the major technological innovations of the Neolithic period was the development of pottery. Clay pots of various shapes and sizes were produced by the technique of coiling snakes of clay in a spiral and smoothing over the joints. The pots were fired either in open fires or, preferably, in a rudimentary kiln consisting of a fire in a pit covered with turf. The pots could be used for storing grain or fats or for carrying water. They could also be used for drinking or eating from and for cooking. Some vessels were ornamented in a variety of ways, either by scratching ornaments into the wet clay or by attaching them to the pot. Here is an early instance of technology serving the need for ornament. It is likely that some ornaments had a ritual meaning, while others were simply regarded as beautiful. Clay pots themselves can be things of beauty and soft clay puts great opportunities and temptation for creativity in the way of the potter. Technological opportunity and human need combine in ways that cannot be disentangled.

We may view the Neolithic revolution either from the point of view of technological determinism or from the point of view of social choices. Neither viewpoint can be proven and neither is totally satisfactory. In my view, we have to regard the social and technological changes as interdependent, as concomitant, as a constellation of circumstances that cannot be separated into cause and effect.

Technological determinism would mean that once the technical possibilities of agricultural production had been discovered and experimentally shown to be feasible, society had to adopt agricultural production and adapt to its requirements and consequences, such as abandoning the nomadic, foraging way of life in favour of settled communities. This deterministic view of the rule of technology over society contains some truth, but it cannot be the full truth. It is true that technology exerts a considerable influence, but it cannot cause society to do things that it does not wish to do. In present-day circumstances, with powerful corporations and powerful public relations and advertising, we may be closer to technological determinism (perhaps commercial determinism would be a more fitting description) than in the days when technology had to win or lose on its merits, without the backing of powerful advocates. In the days of the first tentative steps toward sedentary communities it would seem most plausible that technology opened up new avenues and society was willing to explore and develop these.

It takes generations to develop the full potential of a complex technological system, such as agriculture. In the early days the new technology can do no more than show the practical feasibility of the new ideas and demonstrate the initial advantages. At least some degree of social commitment is necessary for further development to be undertaken. A few individuals must be willing and able to have a go. If the majority of people were unwilling to settle into permanent settlements and abandon their foraging way of life, it would have been difficult to overcome their resistance. Even so, the two types of society, sedentary and nomadic, co-existed for a long period. There were no powerful corporations that had the ability to dictate patterns of consumption; no powerful advertising machinery to convince people of the advantages of one technology over any other. All that the first steps toward cultivated plants and domesticated animals could do was to show the feasibility of producing more food with decreased need to travel over long distances. As this possibility was perceived to give substantial social advantages in the form of increased and more reliable food supplies, society was willing to give it a try. Perhaps they would not have adopted agriculture in the full knowledge of its ultimate consequences, but in the absence of such knowledge the advantages seemed too great to be ignored. Though it is possible that some saw a connection between the adoption of agriculture and sedentariness, the full range of consequences of agriculture certainly was neither foreseen, nor desired, from the outset.

We can imagine the other extreme of the theoretical possibilities, the case in which the technology was developed entirely in response to social desire. We can imagine hunters and gatherers, and more especially their women with infants, becoming tired of trekking all round the countryside, never knowing where they will find shelter or food or water. They may have felt a strong desire to stay in some nice place in a decent hut, near a

[3] Some scholars argue that a few permanent settlements, e.g. Jericho, pre-date the rise of agriculture. See Henry Hodges, p. 30, Technology in the Ancient World, 1970

good supply of water and of basic food, and go out hunting from such a base. But as long as they were totally dependent on gathering wild plants and following roaming herds of wild animals, they had no option but to lead a nomadic life. The desire for a more settled way of life could not be satisfied on command, but may well have played a role in the gradual introduction of domesticated plants and animals. Technology does not simply produce what society orders, but potential inventors certainly keep society's desires in mind.

Neither social nor technological determinism make complete sense, only interplay between technological possibilities and social desire produces technological innovation. Technology and society are interlinked, with neither of them a free agent, an independent variable. When dealing with such complex phenomena as the development of technology and its adoption by society, the separation into cause and effect becomes very problematic and largely futile. Instead, we have to consider a description in which technological invention and social change interact with each other to bring about a change from one socio-technical state to another. And even that is too simple, for everything is in flux and the stable states that we imagine are not very stable and take a long time to achieve. We must think of the Neolithic revolution to have taken many hundreds of years for its completion in any locality. Indeed technical and social change are never completed, all that happens is that there are times of more rapid change that we call revolutions, and times of less rapid change that we call development. And if the development is slow, we call it stability.

It took hundreds of years for sedentariness to become fully established in the sense that the majority of the population had settled in permanent settlements and the accumulation of wealth had led to the establishment of hierarchies and fully-fledged rituals. It was a period in which spectacular feats of engineering produced megalithic monuments and elaborate burial chambers and other feats of construction. So vast are these monuments and so great was the required input of labour that the population of a single village could not possibly have built any of them. They must be the result of regional cooperation, with workers from far and wide participating in the construction and, probably, in the planning and use of the monument. Only cooperation on a large scale or dictatorship over a large region would have been able to achieve constructions of such grandeur.

The case of Neolithic monuments is a very special case of the interaction between technology and society. First and foremost, there must have been a mechanism to express a social desire in a practical form. This mechanism could have been democratic, i.e. some form of regional meetings must have discussed the issue and reached a decision. Alternatively, there may have been a ruler powerful enough to command large resources and take major decisions either single-handed or in concert with a small number of dignitaries. Secondly, there must have been sets of beliefs that required some form of ritual. These beliefs may have included a belief in democracy and a need for regional assemblies, and/or they may have been religious beliefs that became ritualised in some way, as is the wont of religions. One way or another, there must have been a mechanism for reaching a decision on what was required and a further mechanism to negotiate the requirements with the technicians of the day to see what possibilities there were to fulfil the social desires. After all this, a plan for major works must have become established.

The next stage, the stage of execution, required a great deal of organisation, coordination and what we would now call logistics. The right number of people and the right tools and materials had to be available at the right time in the right place. Was that achieved by the ruler or rulers directly, or was there some sort of technical committee or a chief engineer? We cannot know how it was done, but we do know the fact that it was done.

We shall briefly look at one example only - Avebury in southern Britain, built between roughly 3,600 and 2,200 BC. Though this northwestern corner of Europe was among the last to adopt agriculture, it is blessed with some of the most spectacular Neolithic monuments.

The main features of the Avebury site are long mounds (barrows) for burials; enclosures for assemblies and ceremonies (so-called henges); and a very large artificial hill that served no known purpose. The artificial hill, called Silbury Hill, stands nearly 40m high and it is estimated that it took about 18 million man-hours to build. Its purpose is an unsolved mystery, but must have seemed very important to its builders. Perhaps it served spiritual purposes similar to those served later by the great cathedrals, as a communal act of worship, striving to reach into a sphere outside that of living humans.

The second most remarkable structure is a huge henge, i.e. a large circular area surrounded by an earth wall and an inner ditch, with 98 huge sarsens erected along its inner periphery. It is thought to have served needs for ceremonial and assembly. The ditch was up to 10m deep and it is estimated that about 200,000 tons of chalk had to be removed during its excavation. A total of about 200 stones, each weighing up to 50 tons, had to be dragged from a site about 2km away. The stones were probably transported on rollers and erected with the aid of leather ropes and wooden props.

Finally, there are eight long barrows on the Avebury site, the most famous of which is known as West Kennet. It contains five stone chambers at the eastern end; two each side and one at the end of a passage. The chambers contained human bones, though no complete skeletons. A variety of grave goods were found, such as pottery vessels of different periods; flint tools; bone, stone and shell beads; and some animal bones.

The construction of such large earthworks, containing tunnels and chambers supported by timber or stone walls and roofs, is a remarkable feat of civil engineering achieved with stone-age tools. Enormous effort was put into a construction that served purely social and ritual purposes. Technology had become a servant of social desires that went well beyond bare survival, and yet did not appear to serve a commercial purpose, except perhaps as a marketplace. Technology did not provide profits; it served purely social and biological needs. There can be no doubt that these huge monuments were built because of social 'need' and not just because they were technologically feasible. To the contrary, technological and social resources had to be developed, stretched and mobilised to make the project possible. The technology involved is indeed remarkable, but the social aspiration and social organisation involved are more remarkable still. Was it faith that moved the stones?

It is almost certain that some of the rituals involved supernatural, or religious, beliefs, including a belief in an after-life. Perhaps it was thought that the spirits of the dead were able to protect the settlements and had to be kept happy. No sooner was pottery invented than figurines were produced for purposes that cannot be ascertained, but certainly fall into our category of decorative, ritual and symbolic needs. It was not long before bracelets and necklaces and other ornaments were made of stone or clay beads. Neolithic Man attempted to comprehend the meaning of life and death and conjectured some kind of afterlife. And Neolithic Man felt the need for ornament, to bring beauty and pleasure into a life of toil. Improved efficiency of food production made time available for the production of desirable artefacts that had no practical utility.

One of the remarkable features of technology is the way it spreads – diffuses - over vast geographic regions. There are several ways in which diffusion of technology – and of social organisation - can take place. Migration of people is one way, and plenty of migration there was from the very dawn of humankind right into the Neolithic age. Trade is another way in which novelty diffuses. The traders brought with them the latest fashions in tools, artefacts and ornaments and were instrumental in describing new techniques and new ways of life observed elsewhere. Finally, occasional meetings between migrating people, or people settled in a particular region, took place and information was exchanged during such meetings. Toward the end of the Neolithic period, some larger centres were established and people travelled from far and wide to attend ritual or other meetings. Stonehenge and Avebury were such sites in England; other countries had their equivalents. Different researchers emphasize different mechanisms of diffusion and all the mechanisms are likely to have played their part.

During the latter parts of the Neolithic period, experimentation with the extraction, smelting, alloying and use of copper began. At first the metals were rare and were used solely for precious ornaments. Later they started replacing stone in some applications, though it was only when bronze came into use that metal offered a substantial advantage over stone for spearheads, arrowheads, sickles, knives and so forth. The metal ages saw the development of what is known as civilizations, and these are the subjects of the next two chapters.

Main Literary Sources Used for this Chapter (see also bibliography)

Fagan, Brian M. (1999). World Prehistory- a brief introduction (fourth edition). New York: Longman.
Gamble, Clive. (1999). The Palaeolithic Societies of Europe. Cambridge: Cambridge University Press.
Hodges, Henry. (2000). Artifacts- an introduction to early materials and technology (fourth impression of 1989 edition). London: Duckworth.
Hodges, Henry. (1970). Technology in the Ancient World. London: Allen Lane The Penguin Press.
Scarre, Chris. (1998). Exploring Prehistoric Europe. Oxford: Oxford University Press.
Singer, Charles and E. J. Holmyard and A. R. Hall (Ed.). (1954). A History of Technology, vol. I, From Early Times to Fall of Ancient Empires. Oxford: Clarendon Press.
Whittle, Alasdair. (1996). Europe in the Neolithic. Cambridge: Cambridge University Press.

Chapter 2

The Metal Ages

Conventionally, the end of the Neolithic age marks the beginning of the Copper Age, which was the early stage of the Bronze Age and this, in due course, gave way to the Iron Age. For our purpose it is important to understand why, and how, these materials, that gave the ages their names, came to replace stone - and each other - as the principal raw materials for the manufacture of tools, weapons and other artefacts. And, of equal importance, we need to understand how the new technologies related to the substantially altered societies of the metal ages. It was a period that saw the rise of the first so-called civilizations, with their great cities, elaborate social organizations, hierarchies, written documents, and, last but not least, organised armies. Technology began to be used – or abused – to enable societies to go well beyond mere survival and to seek grandeur, conquest, wealth and power.

By the late Neolithic, humans had developed quite sophisticated technologies and, more important, had become keen observers of nature, as witnessed by their agriculture and by their art. The small isolated groups of early Stone Age people had developed into societies. Population density had increased and humans had become largely sedentary, thus establishing close contacts within large groups of people. External contacts were assured by widespread trade and by regional centres of congregation.

The pre-condition for these developments was the establishment of an efficient agriculture, able to feed those who worked in it and producing a surplus to feed those who worked in other occupations. Agriculture thus enabled a society to become stratified in terms of occupations and, of equal importance, in terms of wealth and power. Whereas ownership of land means nothing to hunters and gatherers, the ownership of land became significant with the introduction of agriculture. With the concept of ownership, the twin concepts of wealth and inequality acquired meaning for the first time in history. Leadership might have been meaningful in a society of hunters and gatherers and the rudiments of a division of labour emerged in the late Neolithic. However, an extensive division of labour and social inequalities occurred only in association with advanced agriculture and with real wealth.

Apart from wealth through ownership of land, it gradually became possible during the late Neolithic, and more particularly during the metal ages, to acquire wealth in several other ways. First, directly connected to agriculture, it was possible to accumulate not only land but also farm animals and, at various unhappy stages of history, slaves. Secondly, it became possible to accumulate various artefacts, particularly ornaments. As is the case today, skilfully crafted ornaments, whether or not they are made of precious metals or precious stones, can represent very substantial wealth. Finally, as the construction of buildings advanced and it became feasible to build larger and better equipped and ornamented buildings, the ownership of such buildings, from the simple dwelling house to the luxurious palace, represented wealth.

In general terms, technology created surpluses beyond the immediate essential needs of society and such surpluses could be accumulated in a variety of forms to represent wealth. If the accumulation was unevenly distributed among members of a society, then inequality of wealth arose and an intimate connection between power and wealth was soon established. If we define society simply as a large group of people living in a locality and linked through relationships of work, division of labour, family ties, rules of conduct, and administrative or power structures, then we may say that by the beginning of the metal ages simple societies had become established in many localities. Indeed even in the Neolithic rudimentary forms of society had come into being, sufficiently well organised to construct large civic works that represented substantial communal wealth.

The establishment of society is a pre-condition for rapid advances in knowledge and, thus, in technology. For only society can assure that knowledge is preserved and transmitted. Individuals, even the most talented individuals, discover very little for themselves; their knowledge consists mainly of what they learn from the experience of others. This is why we say that knowledge is social knowledge; society is the keeper of knowledge and knowledge is transmitted from one generation to the next within society.

Modern societies take elaborate care of their knowledge. They have large education systems designed to transmit knowledge systematically to the young generation, and they have many other institutions designed to preserve, increase and transmit knowledge throughout society. These include research institutes and learned societies, libraries, journals and books, and much else besides. But knowledge also spreads via newspapers and other media and by word of mouth. Each of us knows vastly more than he or she discovered directly from experiment or observation. Ancient societies had no organisation designed specifically to enhance and transmit knowledge, but knowledge was nevertheless imparted by parents and elders to the young and spread by word of mouth among adults. The more contact there was between people, the more cooperation between craftsmen and explorers, the more trade with distant communities, the more readily knowledge was accumulated, preserved and transmitted. Without such mechanisms complex technologies, such as the smelting and working of metals, could not have developed and spread.

After countless millennia of social development and development of artefacts made of stone, bone, antlers, leather and wood, not to mention ceramics, humans were ready for experimentation[1] with new materials. Late Stone Age humans examined, and experimented with, all kinds of materials they found in nature. Some of the stones they examined behaved in rather strange ways, very different from ordinary stones that they knew so well. When they found small nuggets of gold, silver or copper, they noticed that these 'stones' did not shatter when hit hard, but changed their shape instead. It became apparent that these materials are malleable, i.e. they can be hammered into desired shapes. Although native copper - copper occurring in nature in pure form - is quite rare, it was found in sufficient quantities to learn how to shape it into useful, or merely attractive, objects. Copper can be hammered into a desired shape, but as its deformation progresses, its malleability decreases. The material becomes progressively harder as it is worked; a phenomenon known as work hardening. Work hardening gives copper the advantage that reasonably hard cutting tools, such as sickles or daggers, can be made from it. On the other hand, continued hammering deforms the material progressively less. This must have baffled early workers, but they found a way out of the problem. If the work-hardened material is heated and cooled, preferably rapidly by quenching in cold water, it becomes malleable again. How this was discovered can only be conjectured. I suppose that people knew that hot copper was soft and assumed that it was worth trying to heat hard copper in the hope that it would turn soft, which it did. That rapid cooling enhances this process of annealing is an added bonus, probably discovered by somebody too impatient to wait for the hot copper to cool down naturally. A suitable sequence of work hardening and annealing makes it possible to produce many desirable copper objects. However, because native copper is rare, it was initially used mainly for the manufacture of precious ornaments and jewellery. The oldest such objects were found in Anatolia and date to about 6,000 BC. The oldest copper weapons and implements were found in Egyptian tombs and date to about 5,000 BC. Silver and gold were also used for the production of jewellery and ornaments from about the same time as copper.

In principle, metals differ from rocks in several important respects. Metals are malleable, meaning that they yield and change their shape under an applied force, such as hammer blows, whereas rocks shatter without having changed their shape. Thus metals can be shaped by the application of force without fracturing them, whereas stones can only be shaped by fracture or grinding. If a metal is subjected to a progressively increasing force, say in a press, it will first deform elastically, which means that it will regain its initial shape when the force is removed. If the force is increased beyond a certain value, the so-called elastic limit, the metal deforms plasti-

[1] We use the word experiment with two somewhat different meanings. When speaking of experiment in ancient times and in some contexts, we mean trial and error. When speaking of experiment in the context of science, we mean scientific experiment.

cally, i.e. permanently. Elastic deformation is useful for the production of springs, and also helps when metals are used to bear loads or blows. Metal implements have a toughness that cannot be matched by implements made of brittle materials. The exact relationship between applied force and deformation of metals was not discovered and formulated till the 1660 by Robert Hooke (1635-1703). Metal technology thrived for two thousand years without a proper theoretical understanding of most of its fundamental properties. Metals generally melt at a lower temperature than stones or ceramics and this circumstance makes it possible to cast metals into desired shapes in moulds made of fired clay. Metal casting was developed quite early in the metal ages and has never ceased to be of crucial importance.

The fact that fundamental knowledge about properties of metals was not discovered till the 17th century need not surprise us. The very idea to explore properties, and seek relationships between facts, did not occur to humans till about the 17th century, with a few exceptions in classical Greece and elsewhere. It is said that Galileo Galilei (1564–1642) was the father of experimental science, whereas men such as René Descartes (1596–1650) and Francis Bacon (1561–1626) laid the philosophical foundations to what we know as scientific method, meaning the quest to interpret natural phenomena on the basis of observation, experiment and rational logical thought instead of on speculation, tradition and theological dogma. Observation of natural phenomena, particularly in astronomy, and their rational interpretation by men such as Nicolaus Copernicus (1473–1543, led to early conflict between science and the Church. By the time Johann Kepler (1571–1630) formulated his mathematical laws for planetary orbits, based on painstaking observations by himself and Tycho Brahe (1546–1601), the Church had more or less come to terms with modern astronomy and the idea that the Earth revolves round the Sun and is not the stationary centre of the Universe[2].

We distinguish between scientific observation and experiment. Observation, coupled with trial and error, led to all early technologies. Particularly agriculture owes its existence to careful observations of nature. Many believe that the siting and construction of megalithic monuments, such as Stonehenge, were based on observations of solar phenomena. Observation, aided by early scientific instruments, led to the first scientific breakthroughs in astronomy and other sciences.

The crucial difference between observation and experiment is that experiment deliberately sets out to test various properties of nature by systematic observations of phenomena deliberately set up. Although some scientific instruments, such as the telescope[3] or the microscope[4] were used for making more accurate observations, later instruments were designed and used for scientific experimentation. In modern times these instruments range from the colossal and ultra-expensive particle accelerators used for experimental observations of fundamental particles; to the huge telescopes and space probes used to observe astronomical and astrophysical phenomena; to quite simple laboratory equipment such as scales, optical microscopes, or distilling apparatus. Many of the scientific instruments have become machines in their own right. We thus use accelerators in medicine, speedometers in cars, and computers for anything and everything.

To establish what now seems to us a simple law, known as Hooke's law stating that strain is proportional to stress, or that metals deform in proportion to the force applied to them, required a great deal of painstaking experimentation. First a specimen of metal had to be prepared, say a round rod, with precisely known dimensions, then a force had to be applied to it, say by clamping one end and hanging weights on the other end, finally the extension of the metallic specimen had to be measured as the load was increased. The procedure had to be repeated many times with the same metal and same kind of specimen, then with different metals, different shapes of specimen and different loads. Not only was the proportionality of stress and strain established, but the elastic limit for different metals was measured as well. No wonder it took till the 17th century to hit upon the idea that useful facts could be gleaned only by experiment, as opposed to trial and error or mere observation. When we consider the complexity of scientific investigation and the amount of effort required to obtain some theoretical knowledge, it should not surprise us that people were unwilling for a long

[2] For popular descriptions of the war between scientific and religious astronomy see e.g. Dava Sobel, Galileo's Daughter, (1999) or Arthur Koestler, The Sleepwalkers, (1959). For a philosophical treatise see e.g. Thomas Kuhn, The Copernican Revolution, (1957).

[3] Developed and used by Isaac Newton and Johannes Kepler for astronomical observations

[4] Developed and used by Antonie van Leeuwenhoek (1632-1723) for microbiological observations

time to spend so much effort to no apparent practical gain. It required a mental sea change before purely scientific experimentation became thinkable and desirable. In ancient Greece science, such as it was, was part of contemplative philosophy. Native copper is too rare to serve as a major engineering material. Much more copious supplies of the metal had to be obtained before its use could become widespread. The discovery of copper ores and the possibility of smelting copper from the ore by heating it in an intense charcoal fire took a very long time, perhaps a thousand years. The relationship between knowledge about native copper and the discovery of smelting copper ores[5] is obscure. My own feeling is that as a result of the success obtained with firing clay and thus obtaining ceramics, people began experimenting with heating all kinds of natural materials. Copper ores are rather brittle rocks, useless for the production of tools, but this did not mean that heating them might not yield interesting results. After all, unfired clay is not a very exciting material either. Before smelting became a practical proposition, clay and ceramics for the construction of furnaces and crucibles had to be available. The availability of ceramic materials was one of the necessary pre-conditions for the smelting of metals. It goes without saying that the discovery of the process of manufacturing charcoal was another pre-condition. Wood-fires cannot reach the temperatures required for smelting. When experiments with heating certain rocks (ores) yielded copper, this was probably recognized as the material already known to be useful and the technique of copper smelting was further developed. It took a great deal of further development by trial and error before the process of smelting was perfected and reasonably pure molten copper was obtained. It is not very likely that the discovery of copper as a material for the manufacture of artefacts was the result of a systematic search for a material that could overcome recognised weaknesses of stone and other known materials. It is much more likely that the introduction of copper was the result of a general quest for practical and useful knowledge, and of fun with experimentation, that had already succeeded in producing ceramics. I think that the discovery of copper and its properties resulted from general exploration, from a quest for greater knowledge of natural materials, though not a systematic exploration that arose with the introduction of science many centuries later. The practical utility of copper was recognised only after some of its properties had been explored and understood.

It is now customary to argue at length about whether a technological innovation resulted from an invention that eventually found a market, so-called technology push, or whether it resulted from a known market demand which technologists were able to satisfy with some novel technology. The distinction contains a grain of truth, but is not very helpful. I think such debates are pretty sterile even under modern circumstances and rather inappropriate for prehistory or ancient history. It is true that some technologies were developed as direct responses and solutions to practical problems. The plough or the wheeled cart are cases in point. Other technologies were developed as a result of general exploration and were found to be useful. Needs in those remote days were so great that it is difficult to think of useless technologies that might have been discovered. It is not very important to distinguish between technologies that were triggered by need and those that were triggered by curiosity; all technologies in those days served very real needs.

The period we are speaking of is one in which technology invariably served needs. The days of technology serving greed were yet far in the future. We might regard two types of technologies as possible exceptions: the technology of warfare and the development of luxury items. Sadly, warfare was perceived as a real need and, even more sadly, this is still very much the case today. Luxury items, such as jewellery, catered for the greed of a very small group of the wealthy and powerful classes. Indeed luxury consumption has always been one of the hallmarks of the wealthy, thus distinguishing them from the rest of humanity.

Technological innovation occurs if a technological possibility is found to be of some practical utility. In other words, technological innovation is the result of the confluence of technological possibility and market demand. The debate about which comes first does not shed much light on the process. In the very early historical periods that we are discussing here, needs were so great and so manifold that it would have been unthinkable for a technology to be other than driven by need. We have earlier established a kind of hierarchy of needs. Top of the list are the vital needs of survival, i.e. food, water, shelter, protection against inclement

[5] Generally speaking ores are mixtures of compounds containing a reasonable amount of the metal in question, such as oxides and sulfides.

weather and predators. The next level of needs are social needs, i.e. needs of social organisation, of assembly and of ritual. These needs arose as early as in neolithic times and found a technological embodiment in the megalithic monuments discussed earlier. In larger and more complex societies, such as in cities, the needs also become more complex, both in terms of housing, of communication, of water supply, of sewage and garbage removal and, last but not least, of food storage and of administration. These needs did not arise till the rise of the first civilizations. With further increases in hierarchical structures and inequality of wealth, the rich required palaces and ornaments. As skills in pottery developed we find some pottery used as ornaments. As skills in metalworking developed, we find that metals, especially precious metals, were increasingly used for the production of ornaments such as jewellery and ornamented weapons. With the development of long-range trade, precious stones, spices and scents arrived on the market and the rich developed an appetite for such luxuries. Technology began to serve the need of the rich for luxuries and thus began to serve greed as well as need. The greed was that of the rich and mighty, not yet the greed of the technologists or the masses, though traders and jewellers profited from the trend.

There was no general cry for new materials; society did not demand metals. Metals were discovered and their practical potential was realised. The considerable amount of exploration and experimentation necessary for the practical application of metal technology required a considerable amount of manpower. There must have been a sufficient number of knowledgeable and skilful people who were able to spend sufficient time and resources on exploration. There must also have been a reservoir of manpower capable of receiving, applying and further developing new knowledge and new skills. Early hunters and gatherers simply could not have spared the manpower for such enterprise. Late Neolithic society had created sufficient surpluses to make such enterprise possible and society had become sufficiently interconnected to provide the cooperation needed for such development.

When copper had become widely available, it became possible to produce reasonably sharp copper blades for sickles and other cutting tools or daggers. Copper proved useful for cutting tools because it was tougher than stone, though not as sharp. Undoubtedly the first call on the new material, especially when it was still very scarce, was the production of ornaments for the ruling classes and the production of arms. Once copper became more abundant, it was found that hammering a copper sheet into a hollow mould could produce copper vessels, to supplement containers made of ceramic, woven baskets or gourds.

There was a lot to be done before copper could become ubiquitous. The ores had to be identified and mined. Charcoal in sufficient quantities had to be produced. Furnaces for the smelting of copper had to be developed. The ore had to be washed to rid it of earthy material mixed with it and then the ore had to be heated to dry it and get rid of some impurities, such as sulphur. The properties of malleability, work hardening and annealing had to be explored and mastered before copper could be effectively shaped into useful objects. Crucibles and moulds had to be developed for the production of cast copper objects. It is clear that the widespread use of copper was dependent on the development of a whole technological system involving much knowledge. This is an early example illustrating the fact that as technology advances, we increasingly have to deal with technological systems, rather than with individual technologies. Several separate techniques and methods become interdependent to form a complex system. It is the system that performs the task, or satisfies the need. We shall meet this phenomenon many more times. The development of the system that might be called copper technology took a long time and it was not till about 3,000 BC that the use of copper became widespread in the Middle East and started to penetrate into Europe. Whereas the production of a stone tool required no more than two stones and a skilled pair of hands, the production of a metal tool required a whole system of interlinked technologies. And, as mentioned earlier, before the system could develop, some pre-conditions had to be met. It is common for the development of a technology to depend upon the existence of other, earlier, technologies.

Although copper proved useful for the production of sheet material and shapes and vessels that could be made from it, it was not hard enough to be truly superior to flint or obsidian for the production of cutting tools. It was superior only in so far as it could be shaped more easily into more complex shapes. Its softness could be overcome by work hardening, and so copper proved useful, but had no overwhelming advantage over stone in cutting tool applications. The widespread use of copper for the production of tools and weapons occurred only when it was found that a small addition of tin produced a superior material – bronze. We do not know what led to the discovery of bronze. Presumably as part of the process of exploration some other ore, tin ore in particular, was added to the copper ore before smelting and the product proved very superior and was further developed. On the other hand, it may have happened through accidental contamination that was reproduced when its benefits came to light.

With the discovery of bronze, a material became available that was hard and strong and could easily be cast or cold-worked into many desired shapes. By varying the amount of tin added to copper, bronze of different qualities could be obtained. The most common bronze in antiquity contained about 10% of tin. The possibilities and techniques of shaping bronze by hammering (cold working) are similar to those of copper. The superior strength of bronze makes it possible to cast very large objects. Hammering can produce strong and sharp work-hardened blades. As bronze is very much stronger and harder than copper, and is shaped nearly as easily, it proved a truly superior material. Thus the Bronze Age was born and bronze spread rapidly from Egypt during the second millennium BC. It arrived in Western Europe in about 2,000 BC and in Britain and Scandinavia about 100 years later. Despite the late arrival of the Bronze Age in Britain, Cornwall, with its large deposits of tin, was destined to play an important role in it. Bronze was particularly suited for the production of daggers and lance tips, but also for the production of chisels. If we consider that almost all the marvellous temples, pyramids and sculptures of old Egypt were produced with bronze as the only material available for the production of stonemason's tools, we must marvel at the amazing skills and ingenuity these craftsmen could muster.

Pure, or nearly pure, iron occurs in nature in the form of meteorites. Meteorites were explored and used almost as early as native copper, but because of their rarity, their use never went beyond the realm of ornaments. The smelting of iron was discovered much later than that of copper. The reason for the delay is easy to see. The smelting of iron ores requires a much higher temperature than the smelting of copper. To achieve such high temperatures, a lot of development of furnace technology was required, including the use of bellows to increase the flow of air through the furnace. The earliest extraction of iron from an ore dates to roughly 2,000 BC and is thought to have started in Anatolia and Persia, where the smelting of copper was well established.

Smelting was achieved by adding iron ore to a bed of red-hot charcoal in either a bowl furnace or a shaft furnace. The bowl furnace, which did not remain in use for very long, was constructed by digging a hole in the ground and blowing air into it through a pipe. The shaft furnace was built of stone above ground. Natural draft was usually supplemented by the use of bellows made of leather. Early furnaces were not capable of achieving temperatures higher than about 1,150°C and the iron obtained from them was a hot semi-solid lump, known as a bloom, rather than liquid iron. The bloom was then re-heated in a furnace and hammered – forged – while red-hot. The forging expelled trapped slag and charcoal and produced the desired shape. This type of iron is known as wrought iron and was the preferred form in the Iron Age. Cast iron, which contains more carbon and is rather brittle, was used to any extent only in China.

The design of furnaces gradually improved and they increased in size, but remained essentially unchanged till about the 15th century A.D. The usual gradual development to larger size and greater efficiency took place. The larger later furnaces had water-driven bellows and the molten slag would run off into pits. The increasing size of blooms is another case in point showing the tendency for a figure of performance to increase over time. Early blooms weighed only about 5kg, whereas by the 15th century blooms weighing over 100kg could be produced in a single firing.

Iron was a truly superior material. Very much stronger and more ductile than bronze, and easily forged into many shapes, iron served well as a universal material for the manufacture of virtually all tools and weapons and many other implements. The only drawback of iron is its vulnerability to rust. This is a great disadvantage that haunts us to this day, but the advantages of iron far outweighed the disadvantages. In effect, the Iron Age began in about 1,200 BC in the Middle East and south-eastern Europe. It became established in the rest of Europe by about 1,000 BC and in China by about 600 BC. Although the knowledge initially diffused from Anatolia and Persia, it soon acquired regional characteristics by local modifications in many localities. We have said that copper technology has to be regarded as a technological system. The same has to be said, with even greater validity, for iron technology, which forms an even more complex system. One of the components of these systems is a body of expertise, experts in identifying and mining metal ores, experts in smelting, and experts in forging iron tools. The trades of miner, iron smelter and blacksmith were born.

Unfortunately, one of the great "advances" of the Iron Age was the production of swords. Stone is obviously unsuitable for such large blades and even bronze swords would be either too heavy or, if light and slender, not strong enough. It is a sad commentary upon human nature that the sword was hailed as the greatest invention of the Iron Age. Some smiths devised a method of obtaining even harder iron – a sort of steel – for the use in blades, especially swords and knives. They heated the almost finished article in a bed of charcoal. Some carbon diffused into the surface layers of the iron and thus converted these into steel. When the hot blade was rapidly quenched, the surface became hardened. It could be further forged and suitably heat-treated (tempered) to give a hard, sharp, yet flexible blade. The best steels during the Roman period came from India, where small blocks of proper steel – not just surface-hardened iron - were produced in crucibles and sold into Europe and the Middle East.

In a very real sense, the Iron Age never ended, for steel is still one of the most important materials used in our society. The dawning of a new technological age does not necessarily imply the demise of the previous technology. Despite to-day's importance of plastics or silicon, steel is still very much with us[6].

Apart from metals, the last few millennia of the pre-Christian period produced many further important technological advances. It would not serve my purpose to describe them in detail, or even to enumerate them all. Instead, I shall briefly discuss those that I regard as most important for my purpose of relating technological advances to social change.

Agricultural production required permanent settlements. Perhaps it is more accurate to say that agricultural production and permanent settlements were dependent upon each other. It is not possible to roam the countryside and have fields and paddocks – these are mutually exclusive ways of life. On the other hand, it is impossible to have permanent settlements and rely on hunting and gathering for food supplies. As soon as permanent settlements were required, ways of building permanent abodes had to be devised. Hunters and gatherers could rely on caves for the winter and temporary flimsy shelters for their wanderings; sedentary populations had to build houses that could withstand the vagaries of the climate. There were not enough caves in the right places, nor many caves large enough to accommodate a whole settlement. Thus buildings were constructed with locally available materials and according to local climatic conditions. Timber was often used, but in places where timber was scarce, stone served as a building material. Some dwellings were half dug into the earth, thus probably offering better protection from wind. Roofs could be covered with reeds, straw, hides, or sods of grass. Houses could be square or round, large or small, with internal divisions or without. All houses had fireplaces and often storage places and stone or timber beds were built in. As settlements grew, lanes had to be constructed between houses. Building techniques advanced continuously and it became possible to build larger buildings of better quality. Only in later times and in larger settlements is there evidence of some form of drainage.

Civil engineering made major strides. Canals were built to transport water over quite long distances to feed irrigation systems. Pipes were made from copper or lead and used to bring water into houses. Even rudimentary sewage systems were built. Elaborate water storage systems were devised and dams were built both for storing water and for controlling floods. Simple roads were constructed and streets were built in settlements. Large settlements acquired defensive walls. Fortresses were built and, as a counterpart, complex machinery for the assault on walls and fortifications.

The horse and the donkey were domesticated and mules were added to the range of domesticated animals suited to be used for riding and as pack animals and, later, for drawing carts or ploughs. The obvious advantage of this form of transportation is that it does not have to rely on any form of road. For the transportation of heavy goods over short distances some form of sledge, the so-called travois, could be used. This consisted of a couple of beams connected by a platform and could be pulled by oxen. Friction was obviously rather great and loads had to be kept low. The obvious thing to do was to put a couple of wheels at the points of contact between the sledge and the earth. But how obvious was this?

The wheel is often regarded as the greatest invention of all time, though it would seem that it was invented many times in different places. One can imagine that when the going got heavy, people would put tree-trunks

[6] See e.g. Arthur Street and William Alexander, Metals in the Service of Man, 1994

under a sledge to reduce friction and get it moving. From the idea of putting loose rollers underneath that had to be brought forward again and again as the vehicle progressed, the idea of attaching something equivalent to a tree trunk might have occurred to many people. Curiously, the earliest known wheels were not simply disks cut off a tree-trunk but were fabricated from three shaped planks clamped together. The wheel had to be attached to an axle and the axle had to rest in a bearing of some sort. Perhaps the really great invention is the bearing, rather than the wheel. For what use is a wheel without a method of attaching it to a platform or frame in such a way that it could carry the load and yet rotate freely? As a minimum, two holes had to be made in vertical surfaces that were part of the vehicle and a fixed axle with a wheel firmly attached to it at each end had to be constructed and inserted in the two holes. In this construction the axle rotated in bearings attached to the cart. Thus the two-wheeled cart was invented and it was a small step from there to a four-wheeled cart for heavier loads. This sounds simple, but it is not. To drive a cart with four wheels round a curve means that the outer wheels have to cover a greater distance than the inner wheels and also that the front wheels have to be pushed sideways into the curve unless they can change their direction, i.e. unless the axle can pivot round a central vertical axis. It was only when each wheel could turn independently on a fixed axle, and the front axle could rotate round a vertical axis, that a truly driveable vehicle was obtained. The road to this development was a long one.

The ancient technique of rolling heavy loads on tree trunks, always placing at the front the one that came out at the end, was a precursor to the wheel. The main difference consists of the fact that the wheel is attached to the load-bearing vehicle, whereas the rollers are not. Solid wheels fabricated by joining together flat planks of wood and shaping them into a circle probably proved stronger than mere disks cut from tree-trunks and avoided the limitations in size that a tree-trunk imposed. However, solid wheels were heavy and this problem of weight was only solved when the spoked wheel was invented around 2000 BC. The production of a spoked wheel is an altogether more complex proposition that required a good deal of development work. The art of the wheelwright came into being.

A large part of the effort of the draught animal was wasted on friction in bearings and friction of the wheels on the road surface. Friction between a dry wooden bearing and a dry wooden axle is great. This problem was eventually solved by producing wheels with iron hubs that turned on a solid axle, either of wood or of iron, and using lubricants to reduce friction further. The spoked wheels were equipped with iron tires, consisting of an iron hoop slipped over the rim of the wheel when hot. On cooling, the iron hoop contracted and held the wheel firmly together. Thus the tire strengthened the wheel and also reduced friction between the wheel and the road surface.

The harnessing of the power of the animal to the cart poses its own difficulties. With oxen, a yoke can be attached to the horns or to the withers of the animal. With horses or donkeys the matter is more complex. They have no horns and the withers are not suitable for holding a yoke. For several centuries a strap was put round the chest of the animal and attached to the cart. This had the severe disadvantage that the strap could slide up and press on the animal's windpipe, thus reducing its traction effort considerably. Gradually the harness was improved by holding it firmly in place so that it could not press on the windpipe, but for heavy loads and heavy horses the solution came only with the invention of the rigid collar, which allowed the horse to exert its full strength in traction. The rigid horse collar is reputed to have been invented in China and did not come into general use in Europe till medieval times.

Why did this development take several centuries? There probably are two parts to the explanation. First, it may have taken a long time to diagnose the problem. It was probably not at all obvious that the horses could have developed greater strength but for the foolish method of harnessing them. And even once the cause was suspected or discovered, the method of solving the problem was by no means obvious. It took a long time to realise that there was a problem, and a further period to find suitable solutions. In this case, the technological innovation was truly driven by a need.

There is plenty of evidence that considerable knowledge of farm animals had been accumulated by the time of the Iron Age. Farm animals were improved by selective breeding and, if the oxen used as draught animals were indeed oxen and not bulls or cows, then castration must also have been invented during this period. This means that the role of testicles in determining a male animal's temperament, as well as its reproductive capacity, had been recognised and acted upon. These discoveries are far from obvious and required powers of observation and abstraction, as well as curiosity and patient experimentation.

It is said that the horse was first domesticated in the middle Dnieper region in the middle of the fifth millennium BC.[7] This seems likely, for the steppe of that region is the place where horses could best develop their potential for speed and strength. Riding in a forest does not bring nearly as much advantage as riding in a steppe. This is a case when natural circumstances caused development to move in a certain direction. The same region also produced covered wagons with solid wooden wheels in the third millennium BC.

Topography and climate and other external environmental circumstances do, of necessity, influence technological developments. It is well known, for example, that the Aztecs, with all their sophistication and their wheeled clay figurines, never developed the wheeled vehicle. Presumably it would have been pretty useless in their mountainous environment and, if that were not enough reason, they had no large domesticated animals that could have been used for drawing such vehicles[8].

The wheel was useful not only for transport but also for the manufacture of pottery - the potter's wheel; for spinning - the spinning wheel; and, in somewhat different form, for grinding grains - the quern. It is perfectly possible that any of these four applications was at the root of the invention in any particular locality and the other applications followed from what we might call cross-fertilisation. An idea that is successful for a particular application can be transferred with relative ease to other applications. It requires the ability to recognise some basic principle – in this case rotary motion – and to grasp that this same principle might be useful for a variety of purposes. Cross-fertilisation, or the transfer of ideas from one sphere to others, always has been one of the ways in which inventions have been made.

I think that as far as transport is concerned, the wheel was a result of experimentation aimed at solving problems arising out of the need to transport heavy goods, such as stones, carcasses, and timber. In its initial stages of development the cart did not prove so very successful because a cart without a road is not very useful for carrying loads. Roads were needed to complete the system of wheeled transportation. The invention of the wheeled vehicle, as so many inventions, found an early application in warfare with the development of a light fighting vehicle; the chariot. Ceremonial types of wheeled vehicles have been found in burial chambers, showing that this technology was used in ceremonial from very early days.

Need was the original driving force that created technology during the early millennia of human existence. Gradually technology became emancipated in the sense that it became able to offer products and processes that were not strictly needed. Technology became able to offer wares that, if all went well, aroused the desire to buy and use them. The offer – an invention – became an innovation if, and only if, some people agreed to use it. Thus invention, need and desire became inextricably bound into a bundle that we call, for the sake of brevity, technological innovation. The inventor changed his/her question from "how can I solve this problem" to the question "if I produce this product, will anybody regard it as useful". A technological innovation occurs if a constellation of technical and socio-economic circumstances is favourable to it.

Innovation in the military field is a somewhat special case because the military are particularly receptive to innovation. Anything that promises to kill enemies more effectively, with less effort and less danger to the own side, is welcome. Anything that promises to protect the own side more effectively from the enemy is equally welcome and the same is true for anything that makes fortifications more effective or, on the other hand, promises to destroy enemy fortifications more efficiently. It should not surprise us that as soon as metals had been discovered and techniques for producing and working them had been developed, metals were employed for the manufacture of arms and armaments. It should not surprise us that as soon as wheeled vehicles were invented a fighting vehicle was developed and as soon as the horse had been tamed it was employed in warfare, both for drawing chariots and for riding. As soon as the bridle, the saddle and, first and foremost the stirrup had been invented, a revolutionary new type of mobile fighting force – the cavalry – came into being and was not replaced till the early 20th century.

---

7 Felipe Fernández-Armesto, Civilizations, (2000), p.113

8 George Basalla, The Evolution of Technology, (1988),pp. 7–11

In heavily wooded or steep terrain both cart and chariot were useless until the time when proper roads were built. Only the open steppe permitted their widespread use and indeed it was from such steppes in Central Asia and the Middle East that chariots and carts originated round about 3,000 BC. The chariot became a mighty fighting machine. It was the military applications that proved the strongest incentive for making the vehicle lighter, faster and more manoeuvrable. Once the chariot was adopted as a weapon, the improvements were obviously called for and may be regarded as social needs that technology was called upon to satisfy. The interplay between technology and need is complex and the lead role may change in the course of a technological development. Sometimes technology leads and offers solutions to problems yet to be articulated, and sometimes a need is articulated and sets a goal for technology to achieve.

The plough is a good case in point. From the beginnings of cultivation, it was recognised that soil had to be loosened and weeds had to be suppressed to create the right conditions for successful sowing. Even before cultivation, edible roots had to be dug up. Both operations could be carried out with a variety of hoes, or with digging sticks. The digging stick could be branched and, sometimes, weighted with a stone. In this form it could be pulled along the field and thus became the forerunner of the plough. In its simplest form and application, a human did the pulling. The digging stick became a plough by the addition of a somewhat broader and harder share, initially of bronze and later of iron. When the stick was replaced by a frame that could be attached to an ox or another draught animal, the transformation to a simple plough was complete. Many more refinements followed over the centuries. The share became bigger, the frame stronger, wood was often replaced by iron. Great improvements came when a so-called moldboard was added that made sure that the plough not only loosened the soil, but also turned it over and thus buried stubble and weeds. Much later, wheels were added that made the work of pulling the plough much less arduous.

One has to suppose that the digging stick and the hoe were invented to answer the immediate needs of digging up edible roots. When the possibilities of cultivation were discovered, these implements were put to the new tasks of loosening the soil and weeding. As plots grew larger, and cultivation more widespread, people began to look for more efficient methods of cultivation and the plough emerged from these attempts to speed up and ease the work to be done. Particularly heavier soils could not be worked without a strong share, a strong frame, and an animal to pull the plough. It is not surprising that heavier soils in the north came under cultivation much later than the lighter soils around the Mediterranean and Middle East. By the stage the plough had developed, a new weak link emerged: friction between the plough and the soil made it hard to pull and this effort was wasted. At this stage the idea of attaching wheels emerged, presumably by cross-fertilisation from wheeled carts and chariots. At each stage of development a new weak link was discovered and remedied. The earliest forms of plough came into use by the end of the fourth millennium BC. Some authors claim that the wheeled plough was invented by the time of the birth of Christ, but others dispute this and think it was much later.

The plough is a good example of a sequence of events that, over the course of time, lead from a simple and rather ineffectual technology to a more complex and more effective one. Improvements are made as weaknesses of a technology are discovered. There is interplay between a need and a technology. The technology improves as it attempts to satisfy the need more effectively. There often comes a stage, however, when the technology leaps ahead of the need and causes all kinds of social change. The plough continued to get bigger until, when the tractor had been invented, it reached a size when only a powerful tractor could draw it. By this stage it had become extremely expensive and extremely efficient, making it necessary to invest much capital and shed much labour. Capital and energy intensive agriculture with very high labour efficiency emerged, and caused all the well-known social and ecological consequences. Technology pushed far ahead of basic need, probably ahead of any reasonable need, and caused as many problems as it solved. This may be a moot point, but it is beyond dispute that advanced agricultural technology is a major cause of social and ecological changes – for good or for ill, probably mostly for ill. Should development have stopped at some stage? Who was to say when? When is a technology good enough to satisfy a need and when does it overshoot the mark and begins to cause more problems than it is worth? These are difficult questions to which there is no general answer. It seems to me that development should stop if it is about to cause adverse environmental effects, and/or if it is likely to cause major adverse disruptions to the social system. When ploughs and tractors became too expensive for small farmers, and when it became apparent that agriculture would have to shed a major part of its labour force, and that the heavy machinery was damaging the soil, and that agricultural surpluses would be produced, it probably would have been time to call a halt to further technical development. This can be seen with hindsight. The real question is whether it could have been foreseen and, if so, whether development could have been halted, and by whom. Generally speaking, it never is possible to halt a technological development that offers substantial benefits to the most influential groups in society.

The dawning of the metal ages marked the beginning of what is known as the early civilizations. Several large cities grew and became the hubs of powerful states. During the fourth millennium BC, civilizations arose mainly along and around river valleys, presumably because of the availability of water for irrigation in these semi-arid regions. The rivers Jordan, Tigris and Euphrates in the Middle East, the river Indus in what are now India and Pakistan, the river Nile in Egypt, and the Huang-Ho river valley in China became centres of civilizations. The first of these civilizations, the Sumerians in southern Mesopotamia, built large cities such as Ur and Lagash. The Babylonian and Assyrian civilizations followed the Sumerians and built cities such as Babylon and Nineveh. The Egyptians had Memphis and Thebes, while the Indus civilization was centred on Harappa and Mahenjo-daro. We shall take a slightly closer look at these civilizations in the next chapter.

Main Literary Sources Used for this Chapter

Agricola, Georgius (1912). De Re Metallica, translated from the first Latin edition of 1556 by Herbert Clark Hoover and Lou Henry Hoover. London: The Mining Magazine.

Fagan, Brian M. (1999). World Prehistory- a brief introduction (fourth edition). New York: Longman.

Hodges, Henry. (2000). Artifacts- an introduction to early materials and technology (fourth impression of 1989 edition). London: Duckworth.

Hodges, Henry. (1970). Technology in the Ancient World. London: Allen Lane The Penguin Press.

Ramelli, Augustini (1620). Schatzkammer Mechanischer Kuenste. Translated into German and published by Henning Gross, Leipzig. See also Ramelli, Agostino. (1976). Various and Ingenious Machines. English translation by Martha Teach Gnudi, Baltimore: Johns Hopkins University Press.

Scarre, Chris. (1998). Exploring Prehistoric Europe. Oxford: Oxford University Press.

Singer, Charles and E. J. Holmyard and A. R. Hall (Ed.). (1954). A History of Technology, vol. I, From Early Times to Fall of Ancient Empires. Oxford: Clarendon Press.

Whittle, Alasdair. (1996). Europe in the Neolithic. Cambridge: Cambridge University Press.

Chapter 3

The Early Civilizations

The major construction works undertaken by local village communities of the late Neolithic imply that they had some form of effective organisation and leadership. The execution of such major works as their great monuments, graves and henges would have been impossible without some form of effective leadership. People obviously also had some religious beliefs, probably in the form of a cult of their ancestors. The incomprehensibility of death and the wish to retain links with the loved departed is a strong stimulant of religious beliefs and associated rituals and cults. The desire to seek protection from natural catastrophes and from enemies provided further incentive for seeking to appease the incomprehensible higher powers. Undoubtedly, such religious cults were part of the social life of neolithic communities. It is possible that an elite was formed from the ranks of the more successful farmers, who had acquired a larger proportion of the available land and produced greater surpluses, thus accumulating some wealth. Religious leaders, shamans of one kind or another, probably claiming abilities to heal the sick and to intercede with the spirits of the ancestors, probably were another part of the leadership. We do not know when martial prowess became a qualification for leadership, but by the time the first civilizations became established, the leading elite certainly included both warriors and priests. The monuments and other constructions of the late Neolithic and the early metal ages demonstrate one of the great strengths of humans: they are able to communicate ideas and plans and are able to cooperate to put the plans into practice. We may regard the great monuments of the Neolithic as monuments to the spirit of cooperation and to human technological ingenuity. Unfortunately the first civilizations also demonstrated the opposite trait of humans: their ability for constructive cooperation is matched by their propensity for destructive aggression. The dichotomy between the will to cooperate within groups and the terrible aggression between groups is the hallmark of primitive humans. In that sense, we have not made much progress.

The organisation of the first civilizations went far beyond that found in the Neolithic. Societies in these civilizations were organised on a much larger scale – not merely at local level – and were much more stratified both in the sense of a strong hierarchy and in the sense of a considerable division of labour. Technology had become too complex to be handled by amateurs. The manufacture of spoked wheels, the manufacture of copper at first, and bronze and iron later, were not only too difficult for the untrained hand, but also organisationally too complex to be managed other than professionally. The same applies to trade, to administration, to accountancy, to water and sewage management, to warfare, and to the building of large elaborate edifices that served various purposes of the state and its rulers.

Some technological, social and geographic pre-conditions had to be met before these civilizations could become established. The first condition was the existence of fixed settlements with a substantial population. The civilizations all had cities as their main centres of population and administration, and cities usually grow out of villages. The second condition was that agriculture had to be sufficiently advanced and sufficiently productive to feed a population substantially greater than the agriculturists alone. Agriculture had to produce sufficient food to feed the workers in the necessary infrastructure of builders, artisans, merchants, administrators, soldiers, priests, and rulers.

Water management was possibly the most crucial technology that made the establishment of large-scale settlements and civilizations possible. The first great civilizations were established in Mesopotamia, where the rivers Tigris and Euphrates served as the sources for an elaborate irrigation system that distributed water from the rivers over a wide agricultural area. To complement the irrigation system, flood controls had to be constructed and in areas of seasonal rainfall the rainwater had to be stored for distribution during the dry season. The water management system varied from region to region and the technologists of the day had to have good knowledge of the rivers and the rainfall, as well as the needs of agriculture. They also had to develop methods of construction for canals, aqueducts and distribution systems. Egypt was exceptional in that it could use the seasonal floodwater of the Nile without too much of a distribution system. The Nile provided both water and fertilizer to a narrow strip of land that formed the basis of Egyptian agriculture and civilization.

Animal breeding became important both for agriculture and for the military. Cavalry and chariots became important instruments of war and needed well-bred horses or asses. Transport, also served by horses, asses and camels was important for trade and for supplying the growing cities with agricultural produce. Good storage facilities had to be built in the cities for the storage and distribution of grain or other food. These storehouses were administered by the state. Plant breeding had to be sufficiently advanced to give good yields and agricultural implements had to be of good quality. Indeed the invention of the plough probably was of critical importance to the establishment of the civilizations.

A further pre-requisite for the establishment of large complex administrations was the invention of a script, even if only as a means of bookkeeping in the first instance, and a small elite able to read and write and do arithmetic. The first accounting system, based on clay tokens, is associated with Ubaid culture and the city of Ur (near the modern An-Nasiriyah in Iraq), precursors to Sumerian culture. Writing proper, so-called cuneiform writing[1], was introduced by what is often regarded as the first of the great civilizations, the Sumerians in southern Mesopotamia. The beginnings of this writing go back to the 4th millennium BC and are associated with the city of Uruk. The Egyptian hieroglyphic script also originated toward the end of the 4th century BC.

The final precondition for the establishment of a civilization is sufficient social cohesion. Society must be willing to comply with certain rules of an organisation that finances and controls public enterprise, collects taxes, settles disputes, maintains civic order, conducts wars, and so forth. Artisans, administrators and soldiers had to be trained, thus some system of training and education needed to be established. It was hardly possible for these states to be completely self-sufficient in all raw materials and manufactured goods, so traders were required to bring all that was necessary from foreign lands. In order to finance such trade, a reverse flow of exported goods was needed.

The term civilization describes a reasonably homogeneous society with common rules of conduct, normally a common language and a common set of beliefs, ruled in some hierarchical organised way, and using a certain standard of technology and style of artefacts. The central authority is based on a city and the urban elite rules the surrounding rural areas. Generally, with a few exceptions such as the Indus civilization, these societies had an organised army. Warfare and weapons became central aspects of almost all civilizations and were often employed for purposes of conquest and pillage. Civilizations usually construct major works of engineering and major buildings, such as temples and palaces, and produce characteristic works of art.

The word civilization is used to describe well-ordered societies with a façade of great architecture and art, no matter how brutal the rulers may be toward their own population and toward the outside world. Indeed we call rulers great if they kill, conquer and loot on a grand scale. My preferred definition of civilization would describe a society where civilized behaviour is the norm, that is non-violent, polite, considerate and law-abiding, within a reasonably just and fair legal framework. Alas, in the ancient world we are speaking of, my preferred kind of civilization had not been invented. A truly civilized society is a society that endeavours to protect its members as far as possible from the hazards and vagaries of life and attempts to provide as secure and as peaceful an existence as possible. A reasonable standard of comfort, hygiene, health care, infrastructure and environmental protection are subsumed under protection from hazards and security. Members of a civilized society should live in harmony with each other and with their natural and human environments. It is painfully obvious that the old civilizations had little in common with what I would like to define as civilized societies. Some contemporary civilizations come a little closer, but still leave a great deal to be desired. In my own view,

---

[1] Cuneiform means wedge-shaped (Latin cuneus=wedge) and the writing is so called because it was written with a wedge-shaped stylus impressed into wet clay.

humankind cannot be regarded as civilized when wars are still commonplace. Aggressive war is, in my view, a most heinous crime. Yet many so-called civilized states, ancient and modern, have indulged in precisely this crime.

By the 5th millennium BC some Mesopotamian settlements had grown into towns of four to five thousand inhabitants. Each town had a temple with an ornamented façade, an offering table and an altar with a statue of a god. In the larger cities these temples had grown into ziggurats; stepped pyramids with the temple built on the flat top. Some of the towns developed into large cities that became centres of power. In the period about 4300 to 3100BC[2] Uruk (modern Warka in Iraq) was the centre of Sumerian culture and power. Uruk had two large temple complexes, each devoted to a different god or goddess. These complexes also served as storehouses for the distribution of food to those who could not grow their own. The precise relationship between priests and rulers is not known; they certainly cooperated closely and might even have been identical. The population of Uruk exceeded 10,000 by about 3000BC and later grew to about 50,000.

The city was sustained by intensive agriculture using a complex irrigation and flood control system and the plough. Livestock played an important role and dates and fish formed parts of the diet. Many craft-based industries flourished, including metalworking, sculpture, engraving, carpentry, shipbuilding, pottery, spinning, weaving and brewing. The lost wax method of casting was invented in the late 4th millennium BC.[3]. Some historians believe that glass was invented in Mesopotamia around 1600 BC, though most sources believe that although glass beads were known in Egypt by 2500 BC, modern glass was developed in Alexandria after about 300 BC. The chariot, especially the four-wheeled chariot, was also invented in this physically and mentally fertile region. Bronze was vital, not only because it was the material from which effective weapons were manufactured, but also because without bronze tools it would not have been possible to produce the elaborate buildings and sculptures of the time. To our modern eyes it seems unbelievable that such complex works of art, architecture and construction were achieved with bronze tools only.

Technology had come a long way in being able to satisfy the essential needs of a large population and to satisfy all the other needs that a civilization required. With the early civilizations and the first large cities the emphasis of technology began to shift from being merely a tool for survival, to becoming also a tool for cementing power, wealth and privilege. In consequence of the warring nature of the various empires, the development of weapons was one of the most important tasks of technology. The bond between rulers and armourers was strong and determined the main thrust of technological development. Weapons and warfare became a major focus of technological endeavour and, alas, have remained so to this day. Transport systems became another major concern of technology, because armies needed transport and so did the large emerging cities for their supplies.

The second major thrust of technology was also determined by demands of the state and the cities. Major public works were required, such as grand palaces as residences for the rulers, grand temples as places of worship and of assembly, statues as symbols of faith and artistic expression, and splendid burials. Improvements in plant cultivation, livestock breeding, water management, and in agricultural implements were necessary to sustain increased populations. City life makes certain demands on technology, such as transportation, storage and distribution of food, housing, administrative buildings, water supplies, drainage and sewerage, roads and streets, and security, including city walls.

Beyond and above that, once hierarchies became established, technology began to serve the special wants of the elites, such as jewellery, ornaments, large houses, luxury clothing and fine furniture or tableware. With the many new products required by the state and by its elites, specialization increased and whole new classes of trained and skilled craftsmen and of traders emerged. The traders brought luxury goods from foreign lands, including spices and aromatics. The work of the stonemasons, sculptors, painters and builders of the Bronze Age stands unsurpassed to this day. The training of the artisans of the early civilizations must have been very

---

2 All dates in this period are uncertain and different scholars use different chronologies.

3 The lost wax method of casting is used for hollow metal castings. A layer of wax is placed between a clay model and an outer clay mold. The wax is melted and the metal is cast into the space left.

thorough and the standard of knowledge and skills very high. The quality and beauty of these antique works of art are breathtaking.

From roughly the beginning of the 3rd millennium BC competition between cities became intense and royal dynasties began to attain power. The cities, such as Uruk, Ur, Lagash, Babylon, and Nineveh were surrounded by massive defensive walls. The palaces and the royal tombs became more opulent and the first mention of slaves is made. Indeed one of the purposes of the frequent wars was to acquire slaves. Gold, silver, seashells and lapis lazuli came into use to ornament sculptures and buildings. This is not the place to tell the tale of shifting fortunes of the various empires, cities, and rulers. The story is extremely complex and is not essential to our main interest.

Even in Sumerian times technology offered real improvements in everyday tools and thus provided the base on which society and its wealth could develop. Wooden ploughs using metal tips were more effective than purely wooden ploughs, and surpassed the efficiency of digging sticks and hoes by a large margin. Irrigation systems increased the efficiency of agriculture, so that large food surpluses, to feed a large non-agricultural population, could be produced. The Sumerian civilization was dominated by city-states, each ruled by a despotic king supposedly appointed by the gods. The cities took it in turn, as it were, to dominate the Sumerian empire. There was a firm mutually advantageous coalition between priests and rulers. Ur had an elaborate tax and bureaucratic system and a written legal code. The rulers lived in great luxury. The kings of Ur were buried in huge graves with their queens and a whole retinue of courtiers. Apparently they had to take poison in order to follow their king into the nether world. Elaborate grave gifts were found, such as sculptures ornamented with lapis lazuli, silver and gold.

The Sumerian army consisted of three different groups: Mobile skirmishers who wore no armour and wielded javelin-like spears and battle-axes. Helmeted infantry arranged in closed ranks and armed with heavy spears. Clumsy chariots, pulled by donkeys, supported the foot soldiers. The bow came into widespread military use only after about 1500 BC.

After the fall of Ur the Sumerian civilization declined and Babylon (near the modern town al-Hillah, Iraq) became the major city in Mesopotamia. A dynasty was established in 1894 BC and the best-known king of this dynasty, Hammurabi (1792 to 1750), conquered a wide territory and raised Babylonia to the leading power of the day. During the rule of Hammurabi many temples were built and the network of irrigation canals was extended. A written code of law, the Hammurabi Law Code, was promulgated. It was engraved on a great stela (standing stone slab) in cuneiform script and is one of the very earliest written collections of laws. An interesting feature is that it distinguishes between three classes of inhabitants: slaves; royal employees and retainers; and free land-owning citizens. Babylon was sacked in 1595 BC by the Hittite king Mursilis.

A new major actor arose when the Assyrians gained independence from Mitanni in about 1340 BC. In frequent wars, chiefly against Mitanni and Babylonia, the Assyrian empire gained importance, temporarily holding most of Mesopotamia. The rivalry between Assyria and Babylonia continued, though both came under pressure from Aramaean and Chaldean tribesmen, pushing in from Syria. Eventually the Assyrian king Adadnirai II (911 to 891) gained the upper hand. His successors moved the Assyrian capital from Ashur to Nineveh (near modern Mosul, Iraq) and expanded Assyrian power.

The Assyrians had a professional standing army, backed up by a citizen army. Their king Ashurnasirpal II (883 to 859) introduced cavalry into the army as an addition to the chariot force and the infantry. Although many armies introduced cavalry, the horse and rider did not become fully effective until the stirrup was invented in the steppes of central Asia in about the 2nd century BC. The cavalry and chariots became the main attacking force, relying on speed and on missile power. The infantry formed the second wave, using bows and spears and wearing helmets and chest armour. Originally separate shield bearers protected the infantrymen with huge shields. Presumably because of problems of coordination of movement, the shields became smaller and the separate shield bearers were abolished. It is believed that the Assyrian army numbered up to 60,000 infantry, 1,200 cavalry and 4,000 chariots. All winners treated the losers badly, but the Assyrians apparently surpassed all their predecessors in unprecedented cruelty. In 689 the Assyrians destroyed Babylon, but their luck ran out and some 80 years later the Assyrian empire collapsed and their capital, Nineveh, fell in 612 BC to Persian and Babylonian forces.

Babylon experienced a revival and regained considerable importance after a Chaldean leader made it the capital of a new kingdom that again became the dominant power in Mesopotamia under king Nebuchadrezzar II (or Nebuchadnezzar), who ruled from 605 to 561. His claim to fame includes the construction of the hanging gardens of Babylon, one of the Seven Wonders of the World. The gardens formed part of the royal palace and were splendid roof gardens laid out in a series of ziggurat terraces. Water was pumped up to the gardens from the river Euphrates. The centre of the city was marked by a great temple dedicated to the god Marduk and by a 90 m high ziggurat (known as the Tower of Babel) with a temple covered in blue glazed tiles on top. Nebuchadnezzar is also famous for building additional city walls and a ceremonial processional way, paved with stone, that served mainly the transportation of religious statues between temples. This was probably the first paved road ever built. He also built an unpaved road connecting the Persian Gulf with the Mediterranean Sea, a distance of about 2,400km. Last but not least, Nebuchadnezzar acquired biblical fame for capturing Jerusalem in 597, and again in 587, and deporting many of the inhabitants of the kingdom of Judah to Babylon.

One of the major powers of the Middle East was Egypt, with its various kingdoms spanning a period from 3000 to 300 BC. The Sahara became a desert during the 4th millennium BC and this forced its population to migrate to the fertile Nile valley. Egyptian agriculture was unique among the river valley cultures in that it used the annual flooding of the river Nile to irrigate and fertilize its fields. Only a few canals needed to be dug to distribute the floodwaters. The Nile normally flooded in August and the water subsided in the autumn, allowing the crops to grow in the mild winter and be harvested in the spring. What was regarded as surplus grain was collected as tax and stored in state owned storehouses for distribution to the priests, administrators and craftsmen. Some reserves were kept for lean years when the flood failed. The first ruler of the unified kingdom of Upper and Lower Egypt was Narmer, c. 3000 BC. The city of Memphis (near modern Cairo) was founded and the principle of theocratic kingship became established during this early dynastic period that lasted till 2649. It was the period when copper and bronze were first introduced and writing came into use. Egypt had two scripts, the hieroglyphic (pictorial) script used for inscriptions carved in stone, and the hieratic script[4], used for writing with ink on papyrus.

The next period, 2649 to 2134 is known as the Old Kingdom. The most famous pyramids were built as burials for kings during the Old Kingdom. King Djoser (2630-2611) had his stepped pyramid built at Saqqara. It was the first pyramid built entirely of stone and was designed by the first known architect in history, Imhotep, who was later deified. The smooth pyramids of Giza were built for Cheops (Kufu) (2551 to 2528), his son Khephren (2520 to 2494) and for Menkaure, Khufu's successor[5]. Many other pyramids were built during many centuries, but the Cheops pyramid, at 146 metres, is the highest and most impressive of them all. The pyramids represent the most amazing feat of stonemasonry, building and organization. Experts have not agreed to this day how the buildings were accomplished, but one of the more plausible theories is that the huge stones were transported on sledges, and perhaps rollers, up temporary ramps built alongside the rising pyramid. It remains a mystery how the stonemasons achieved the precision that allowed the stones to fit together so perfectly. Another truly amazing fact about these huge marvellous buildings is that they did not serve any practical purpose. After all, royal graves do not really need such edifices. They were built for purely religious cum political reasons, as acts of reverence, faith and worship, and as symbols of power. Technology had moved a long way from serving only needs of survival and had become the servant of religion and of the high and mighty.

The Old Kingdom collapsed after a succession of low floods in about 2150, and with it ended the building of pyramids. After an intermediate period the Middle Kingdom emerged (2040 to c.1640). The Middle Kingdom became more expansionist and aggressive. It collapsed c.1640 and during a second intermediate period the Hyksos (who came from Palestine) conquered Lower Egypt. They brought with them much technology from the Near East, such as wheeled vehicles, and various bronze tools and weapons. They introduced the war chariot to Egypt and a very effective composite bow, made of layers of different woods, as well as the scimitar. The felloes of wheels for chariots were produced from bent ash, showing that the art of bending wood had been

[4] After 660 BC a simpler demotic script largely replaced the hieratic script.

[5] There are considerable differences in the modern spelling of ancient names. Khefren, for example, is sometimes spelt Khafre and Kufu is sometimes Khufu. Cheops is the Greek spelling.

mastered. The hub was usually made of elm and the spokes of oak. The axles, however, remained fixed for quite some time and impeded the mobility of the chariot. The Hyksos also brought new types of crop, new musical instruments and new domestic animals, including the horse. Eventually the Hyksos were expelled and the New Kingdom arose and became the greatest power in the Middle East. The title pharaoh was introduced and Egypt achieved its greatest extent under pharaoh Thutmoses I (1504 to 1492), when it conquered the whole Levant (Middle East) up to the upper Euphrates. In the south, it reached to the 4th Nile cataract, thus including much of Nubia with its gold and its trade relations with the rest of Africa, bringing in slaves, ebony, and ivory. The Levant provided a useful buffer zone against the powerful empire of the Mittani. Thebes (modern Luxor), with its major temples of Karnak and of Luxor and its necropolis on the west bank of the Nile, became the capital from which the whole of Egypt was ruled.

The mighty Egyptian army used two-wheeled chariots drawn by a pair of horses, manned by an armoured driver and an archer. The infantry was equipped with lances, shields and helmets. The bow and arrow, personal armour, javelins, battle-axes, and swords were all developed and improved during this period.

The earliest fully recorded battle, though certainly not the first battle ever fought, is the battle of Megiddo (in present day Israel), in the late 15th century BC. Here the Egyptians under Thutmoses III successfully fought an assortment of enemies who had formed an alliance against them. The Egyptians advanced further north and eventually had to fight the Hittites, whose empire was centred on Anatolia. In the battle of Kadesh (in modern Syria) of 1282 BC, the Egyptians were led by Ramesses II and the Hittites by their king Mutwatallish. Both armies made extensive use of chariots, followed by tightly packed infantry equipped with shields, spears and curved slashing swords. Bowmen followed behind the spearmen. Both armies claimed victory and the war ended with the signature of a non-aggression treaty.

After this brief excursion to Egypt, we return to the region of Mesopotamia. The wheel of fortune turned again when in 550 BC Cyrus the Great became king of Persia and made it the dominant regional power. The Persians conquered most of Lydia (in Western Anatolia) and captured their king Croesus, famed for his wealth, in 546 BC. After Lydia, Cyrus turned his attention to Babylon and captured it in 539 BC. Under king Darius (521 to 486) the Persians conquered Egypt and expanded their empire as far as the Indus and the Caspian Sea, thus turning Persia into the largest ever empire. After these conquests the Persians turned their attention to Europe and the famous wars with the Greeks began. Originally the Persian army relied on moderately armoured infantry equipped with spears and bows. Later they improved the army's mobility by strengthening the cavalry and increasing the number of chariots. The cavalry and the chariots became the spearhead of the Persian army, with the infantry supporting them with showers of arrows. The Greeks used entirely different tactics and fought with deep closed ranks (the phalanx) of so-called hoplites, infantrymen wearing bronze armour, using a shield known as a "hoplon" and armed with iron-tipped spears. Some archers and slingers were mixed in with the infantry and the relatively small cavalry was used on the flanks only.

Before returning to the Persian-Greek wars, we need to say a few words at least about Greece, considered by many as the cradle of Western thought. The collapse of the Mycenaeans left Greece in some turmoil and stability was not re-established till the 10th century BC. There were two main population groups: the Dorians who had migrated from the Balkans and the Ionians who were the direct descendants of the Mycenaeans. Throughout the 9th century, agricultural techniques improved, the population increased, trade expanded and prosperity grew. The Greek alphabet was developed during the Archaic period (800 to 480). It was an expanded and adapted version of the North Semitic alphabet that came to Greece via the Phoenicians, hence it is often referred to as the Phoenician alphabet. Power in Greece was based on cities, of which two, Athens and Sparta, became dominant. Sparta had a particularly strong army, including the only professional standing army, and a severe education system for boys – hence the term Spartan. Gymnastics and warfare were the main subjects taught. Athens was more of a centre of philosophy, literature and learning and invented and experimented with various forms of democracy. Greek cities were often at war with each other, all using closed ranks of hoplites advancing upon each other, killing and pushing till one side submitted. Athens depended on its sea trading routes and its main military strength was based on its navy.

The warships of the time differed in size and in the number of tiers of oarsmen. The smallest vessel had 25 rowers on each side and was used for exploring and raiding. A similar ship, though more suited to fighting, was the unireme. It had the same arrangement of rowers, but had a reinforced prow and was usually equipped with a beak or ram. The next larger ship had two tiers of rowers arranged above each other and was known as the bireme. It had been developed by the Phoenicians and by the 5th century had been largely replaced by the trireme. The trireme, the mainstay of the Athenian navy, was usually 35 m long, had three decks of rowers and carried a crew of 200. The beak was of bronze and was designed to pierce the enemy ship below the water line. The naval tactic was to attempt to hole the enemy ship, though shearing off the oars was an alternative way of crippling a vessel. Sometimes grappling and boarding an enemy vessel and shooting officers and steersmen with arrows was attempted. Triremes could sail, but in combat they used the speed and manœuvrability of trained rowers.

Merchant ships were invariably sailing ships that could make 4.6 knots with the wind and 2.1 knots against it. They gradually increased in size and by the end of the 4th century BC could commonly carry loads of 150 tons with some up to 500 and more tons. Steering was by a pair of oars – the stern-post rudder was not invented till the 13th century AD.

During what is known as the Classical Age, Greek technological innovations included wine and olive presses, both in the form of lever presses or screw presses. Greek roads were neither surfaced nor drained and pack animals were more important in land transport than wheeled vehicles. An ingenious method of carrying heavy building beams was to use them as axles for large drums and roll them along. Another Greek invention was the crane with multiple pulleys, presumably used for loading or unloading ships and for building. Stone columns used wooden pins in sockets to locate the parts of the columns. It is hardly necessary to mention the well-known impressive temples, palaces and fortresses built by the Greeks and decorated with wonderful figures carved in relief. Water pumps of different design formed part of the Greek technological arsenal. They invented both the twin-cylinder force pump and the famous Archimedes screw that was widely used for pumping water.

Greek philosophy, political thought and drama undoubtedly form the basis of much of our own thinking. Greek science, on the other hand, is not a direct precursor to our empirical approach to science. Though the Greeks did observe natural phenomena with some care, they did not believe in testing their scientific theories against their observations. The two aspects were kept in separate compartments, with purely speculative theories being given higher value than mere observation. Nevertheless, some of the achievements of men such as Archimedes (c.290-280 to c.212) or Pythagoras (580 to 500) have stood the test of time. It goes without saying that Greek sculpture, Greek architecture, and Greek pottery were all outstanding and have formed much of our artistic taste. The artistic excellence of their art was matched by the technical perfection of their works.

When Persia had run out of places to conquer, it used an uprising in an Anatolian Greek colony, that had been occupied by Persia, as an excuse to turn their attention to Europe, ostensibly to punish the Greeks for this uprising. The main problem of the Persians was to keep their large army supplied. The only solution to this logistic nightmare was to have a fleet of supply ships follow the army along the Mediterranean coast. The Persian army probably consisted of some 7,000 combat troops, including about 1,000 cavalry and some chariots, and a large number of auxiliaries, including the oarsmen[6]. The foot soldiers were equipped with bows, and the cavalry was armed with javelins, bows, or lances. The Persians chose Marathon, some 40 km (25 miles) from Athens, to do battle with the Athenians in 490 BC. Surprisingly, the Athenians routed the mighty Persian army and are said to have killed 6,400 Persians, with only 192 Athenian dead. Legend has it that an Athenian messenger ran all the way to Athens to relay the victory before dropping dead. This is the legendary basis of the modern marathon run.

The war was not over yet. Darius died in 486 and was followed by Xerxes, who in 480 led a new expedition to Greece. His army is said to have numbered some 40 to 50,000 men in total, presumably not all combat troops. The army marched and a naval force protected their flank and their supply ships. Many Greek cities surrendered, but a Spartan force, led by their king Leonidas, held off the Persian army at Thermopylae for several days. The Persians managed to outflank them and marched on Athens, which they sacked. Leonidas and a force of

6 Some historical sources describe the Persian army as very much larger, but modern historians view these figures with some scepticism.

some 300 Spartans, augmented with ancillary forces, delayed the main Persian army long enough to enable the main Spartan force to escape and regroup. Another legendary feat, with all 300 Spartans killed. The Persian fleet was lured by the Athenians into waters near the island of Salamis, where it was almost totally destroyed in the battle of Salamis by the Athenian and Corinthian navies. Much of the Persian army and the remaining fleet withdrew to Asia, though quite a large force remained near Plataea. In 479 this army was heavily defeated by a Hellenic force, led by Sparta. A further defeat at Mycale completed the rout and drove the Persians out of Europe.

Though the rise of Rome falls into the end of the period we are discussing, it is such a prime example of an expansionary civilization that I cannot resist the temptation of mentioning it, albeit very briefly. The really important achievements of Rome lie in the realms of administration, politics, law and literature, but Rome also used and developed technology extensively in support of its huge armies and very advanced civilization. Technology served two major purposes of Roman society: military conquest and the technological underpinning of civic society. Military conquest required arms and armaments, fortresses and siege machinery. It also required what we now call a logistic infrastructure of roads and transport and the handling and transmission of information. The underpinning of civic society required great monumental buildings as well as the provision of clean water, public baths with good heating facilities, and a high degree of domestic comfort for leading citizens. The usual problems of supplying a major city with food required the solution of problems of transport and distribution, not to mention the need to produce sufficient food in the first instance. In modern parlance we would say that Rome supported a large arms industry and major civil and building engineering projects, such as the famous baths, circuses, roads and aqueducts.

Military technology served the Roman desire to expand to an enormous size and to hold the far-flung empire together by military and administrative means. The formidable Roman army consisted of legionaries as the elite foot soldiers, and ancillary specialist troops, such as cavalry, archers and slingers. All soldiers wore body armour of one kind or another. Some, especially ancillary troops, wore chain mail, though in later times plate armour prevailed because of its greater effectiveness. The Romans designed a form of articulated plate armour – lorica segmentata – that replaced the solid plate armour used earlier. It consisted of segments of armour, joined together by leather straps. It afforded excellent protection and yet allowed considerable freedom of movement. Roman armourers were able to produce sophisticated iron products in great numbers. The legionaries carried short swords with hardened blades and a javelin. The wooden shaft of the javelin was tipped with a soft metal shank with a hard metal tip. The idea was that the sharp metal tip would pierce its target and the soft metal shank would bend, making it impossible to throw the javelin back at the Romans. The soldiers also carried shields, often made of laminated wood, rather like modern plywood.

The army was largely self-sufficient in its technology. Military commanders were involved in technological developments and armourers and other artisans were often part of the army itself. In occupied territories the army built a closely spaced network of legionary fortresses. These accommodated the soldiers and their officers, as well as workshops, storage facilities, a hospital, and bathhouses. The fortresses were first built quickly as temporary structures from timber, and later constructed in stone. They were substantial self-contained units with a standard layout of fortifications, buildings and streets. Water supplies and drainage were provided in accordance with local circumstances. The total investment in labour and materials into such fortresses was enormous, but the purpose of housing the army and having it at hand to counter or deter any possible external attack or internal rebellion was apparently of supreme importance[7].

Another aspect of Roman military technology is concerned with siege. They built several different models of devices that could shoot sharp metal bolts over a fairly long distance – the ballista. It was something like a stationary version of a crossbow, itself a Roman development. They also produced a variety of catapults and battering rams, as well as siege towers and mobile shelters to protect the attacking soldiers from missiles hurled at them by the defenders[8].

---

7 See e.g. Shirley, Elizabeth. (2001). Building a Roman Legionary Fortress.

8 See e.g. Oakeshott, R. Ewart. (1960). The Archaeology of Weapons: arms and armour from prehistory to the Age of Chivalry. London: Lutterworth Press.

To give the army its required mobility and flexibility, and to give the empire its political and administrative cohesion, a good system of transport and communications was an essential requirement. The Romans constructed a dense network of roads, connecting all the major towns and fortresses. Whereas in earlier times roads were just simple cart tracks built where necessary without any overall plan or system, the Romans went about the business of road building with an unprecedented thoroughness. Roads were of varying width, from 4 to 9m, depending on their importance. They were all layered: a foundation of large stones, then a layer of smaller stones, followed by a layer of gravel. For minor roads the gravel was rammed to produce a reasonable surface. For major roads a further layer of large flat stones – the pavimentum – formed the actual road surface. The real secret of the Roman road was drainage. The foundation drained well and the surface was cambered. Ditches on both sides carried away the run-off. The Roman civil engineers realised that mud was the greatest enemy of transport and drainage was the only means to beat mud.

Rome wanted to rule as much of the world as possible, presumably in order to gain economic advantage, to achieve security from potential raiders, and to bestow the benefits of Roman civilization upon a barbarian world. No full answer to the question why so many countries, from ancient times to our day, have attempted to build empires by conquering other countries. The answer I have given is lamentably incomplete, but this book is about technology and not about imperialism, though a link between them is undeniable. Technology has often been developed and misused in support of imperialist aspirations, and is so misused to this day. To make conquests and build the Roman Empire, a strong army was the most essential requirement. To maintain the huge empire, it was essential to build an effective administration and effective communications. Finally, to bestow civilization, required urbanization and the building up of civic amenities.

Roman civilization was a well-ordered political system, i.e. a system of decision-making and distribution of power and wealth. It was also a highly sophisticated legal system, with a codex of law that is still regarded as exemplary in many parts of the world. The Latin language became the lingua franca; the Latin alphabet came into widespread use. Literature developed both in the sense of poetry and in the sense of writing history and science. All these developments required an efficient agriculture, capable of feeding a large urban population, and the development of urban centres where all these civilised activities could take place. In fact the Romans built a large number of planned cities with administrative buildings, courts of law, markets, dwellings and shops and a grid of streets. Much of the building was in timber and posed a fire hazard, hence the Romans introduced the first organized fire fighters, equipped with water containers and pressure pumps mounted on carts. The wealthy classes tended to invest in land and built elaborate villas in the country. These villas provided separate rooms for separate activities, unlike earlier dwellings that consisted of a single room. The villas were often beautifully ornamented with mosaic floors and wall paintings. They usually had a private bath with underfloor heating, a precursor to modern central heating systems. The towns required water supplies and sometimes the water had to be brought over long distances by aqueducts. Drainage and sewerage had to be provided and was carried in stone-clad and covered ditches or in ceramic pipes. Water was often carried and stored in lead (plumbum) pipes and containers; hence the word plumber.

Roman society pursued its main aims with singular determination. The main aim was the growth of the empire to achieve wealth and security, though perhaps it became an obsession and an aim in its own right, without consideration of the benefits or disadvantages. Romans believed that large and powerful was beautiful. Wealth was gained by favourable imports and trade, as well as by carrying off slaves and loot in wars. The state was all powerful and regulated most things. Rome is an example of a determined and autocratic state. In a society that knows exactly what it does and does not want, technology may have some autonomy, but is basically subservient to the general aims of society. It is true to say that technology invents, but it invents in view of circumstances where society chooses the technologies it is willing to accept within narrowly prescribed social goals. Roman technology had the task of strengthening the empire and making life comfortable for the ruling elite. Strengthening of the empire required mostly military technology and an infrastructure of communications and buildings. Strengthening the empire also required strong administration and keeping the bulk of the population quiet by providing circuses and many urban amenities for them.

By this time in the history of mankind, technology had come a long way from merely enabling humans to survive. It was now supporting social requirements that can be characterised as dominated by greed, lust for power, and a desire for order in all aspects of life. Technology provided great wealth and luxury for the elite and provided the means of achieving wider social goals, set by the elite. Technology also provided the means for feeding a large population and providing their basic needs at a much higher level of comfort than in the Stone Age. Finally, technology provided the weaponry and the logistics for large armies. Technology had not, however, become a main source of wealth, nor the prime mover in the quest for more private wealth. The main sources of wealth were agriculture, conquest and trade, whereas technology remained an ancillary instrument for the achievement of social goals.

The ruling elite made demands on technology in accord with its own perception of what society needs. The demands included the satisfaction of the most elementary needs, such as food supplies, clean water, basic housing, and so forth. Elementary needs by this time in history are not quite as basic as during the Stone Age, but basic still. The state may pay for some of these requirements, such as irrigation and water management systems, directly out of taxation[9]. Other needs, such as food and perhaps housing, will be paid for, or produced, by the consumer and the balance between demand and supply will be found by a rudimentary barter mechanism. The elite may also decide, and mostly did decide, that expansion of the state is a legitimate and essential social need and that the state should build up and maintain a military force to achieve this expansion. A further social need, as perceived by the ruling elite, consists of palaces, monuments, splendid graves, temples, buildings for public administration, storehouses, and so forth. These were all social needs articulated by the elite and went far beyond the essential needs of survival. The need for expansion, domination and conquest was, of course, driven predominantly by greed.

The elites had their own private needs, above and beyond the basic needs for food and shelter. They required, or, more accurately, desired a whole range of luxury goods. I do not think that they were able to articulate these wishes, except in very general terms. The items for luxury consumption were suggested to the elite by their suppliers, and these were either merchants or artisans. The elite as private consumers could chose between the items on offer, be it imported luxuries or luxuries produced by local craftsmen. Even in those days the craftsmen, i.e. the technologists, made offers to the consumers and tried to arouse their interest, rather than being told by the consumers what their needs were. Luxury consumption was, of course, driven by greed – not by need.

Military expenditure pursued two aims: defence against outside raiders, and offence with the aim of conquest and plunder. The desire for conquest was driven by greed – by the desire to own slaves, gold, silver, jewellery, and precious stones, that could all be obtained by plunder in conquered lands. It was also driven by vanity and the lust for more power, more grandeur, more vassals, and more fame. Military commanders were drawn from the governing elite and military technology was under the direct control of such commanders. Local artisans, particularly armourers, were attached directly to the Roman military. In the supply of arms there probably was real interaction between the articulation of needs and the capabilities of technologists.

Some form of religion played a major role in all the civilizations and priests became members of the privileged classes, often at the very top of the social pyramid. Indeed religion was often used as the spiritual argument for civil obedience and helped to cement the worldly hierarchy by appealing to supernatural powers. Religion provided the simple uneducated people with a prescription on how to deal with dangers and disasters. They would pray for rain, pray for the sick, and pray to be spared from floods, plunder and pestilence. And they gladly paid the priests to serve as intermediaries between them and the superior mysterious divine powers. The priests regulated their everyday lives and they believed that by obeying the rules they would be spared the dreadful consequences of disobedience and would lead a decent and orderly life in this world and in the next. The priest became surrogate fathers, guides and protectors.

The priests provided the spiritual underpinning for the rule of the elites. Technology provided the material underpinning in the shape of military technology and of worldly splendour of the rulers. A ruler who lived in worldly splendour and claimed divine powers commanded greater respect and more obedience than rulers

9 Taxation may, of course, have taken many forms, such as a proportion of the harvest or other products, forced labour or military service or, in later times, money.

without these trappings of power. Technology had come a long way since the time when it provided only the barest necessities of life.

The Main Literary Sources Used for this Chapter

Fernández-Armesto, Felipe. (2000). Civilizations. Basingstoke: Macmillan.
Haywood, John. (1997). Ancient Civilizations of the Near East and Mediterranean. London: Cassell.
Hornblower, Simon and Anthony Spawforth (editors). (1998). The Oxford Companion to Classical Civilization. Oxford: Oxford University Press.
Oakeshott, R. Ewart. (1960). The Archaeology of Weapons – Arms and Armour from Prehistory to the Age of Chivalry. London: Lutterworth Press.
Pareti, Luigi with Paolo Brezzi and Luciano Petech. (1965). History of Mankind, vol. II, The Ancient World. London: George Allen and Unwin.
Shirley, Elizabeth. (2001). Building a Roman Legionary Fortress. Stroud: Tempus Publishing Ltd.
Singer, Charles and E. J. Holmyard and A. R. Hall (Ed.). (1954). A History of Technology, vol. I, From Early Times to Fall of Ancient Empires. Oxford: Clarendon Press.
Singer, Charles and E. J. Holmyard and A. R. Hall and Trevor I. Williams (Ed.). (1956). A History of Technology, vol. II, The Mediterranean Civilizations and the Middle Ages. Oxford: Clarendon Press.
The Times History of War. (2000). London: Harper Collins.
White, K.D. (1984). Greek and Roman Technology. London: Thames and Hudson.
Wilkinson, Philip. (2000). What the Romans Did for Us. London: Boxtree.

Chapter 4

The Middle Ages

Of the numerous definitions proposed for the Middle Ages, we shall adopt one of the currently more common ones, defining the Middle Ages as the period between the coronation of the Frankish king Charlemagne as western Roman Emperor in 800AD to the fall of Constantinople to the Ottoman Turks in 1453. The fall of Constantinople to the Muslims had grave consequences for Europe. On the one hand, it removed the Mediterranean from the European zone of influence; on the other hand it brought many classical scholars from Constantinople to Italy and thus contributed to the Renaissance movement. Indeed some scholars regard the Renaissance movement, the attempt to resurrect classical Greek and Roman culture as a secular enterprise, as a defining moment of the modern period. The date 800 AD is chosen as the beginning of the Middle Ages, because the crowning of Charlemagne as Roman emperor ended a period of about 300 years when the western Roman Empire did not exist and the only Roman emperor was the eastern, Byzantine, emperor. Some scholars date the Middle Ages from the fall of Rome to the Goths in 476 AD. The period from the fall of Rome to the crowning of Charlemagne as Roman emperor, i.e. 476 to 800, is then regarded as the earliest part of the Middle Ages, generally known as the Dark Ages. It was a period of great turbulence, when many Germanic tribes were on the move and conquered various parts of the dying Roman Empire. Britain was invaded by several tribes, collectively known as the Anglo-Saxons. The period is of no interest to us, as no stable civilisation was established and technology did not advance. Our discussion will concentrate mostly on the late Middle Ages and even a little beyond – when technological, economic and social developments had reached a relatively stable state, which may be considered to have laid the foundations to the next state – the rise of capitalism and the industrialisation of production.

To regard the Renaissance as the end of the Middle Ages makes a great deal of sense. Throughout the Middle Ages proper, not counting the dark ages, the church ruled supreme over all intellectual enterprise. Almost all learning was in the hands of the church or, at the very least, under the strict control of the church. Thus intellectual activity was confined to the interpretation and re-interpretation of church doctrine. Science in the sense of observing natural phenomena, or carrying out deliberate experiments, and attempting to interpret the results in a framework of scientific theories was not possible in the dogmatic mind-set dominant throughout the Middle Ages.

It is often said that technology stagnated during the Middle Ages. This is not quite correct, though it is true that no fundamentally new developments occurred. On the other hand, the craft-based technologies of the day were constantly improved and developed to a high degree of perfection. Technology developed in small incremental steps, mainly on the basis of improvements made by individual craftsmen. Sometimes the improvements were driven by a desire for perfection, sometimes by practical problems that needed solutions. There was no theoretical framework to guide technology and indeed its organisation was, like the rest of society, extremely conservative.

Throughout the Middle Ages, the strict adherence to faith as taught by the official church seemed irreconcilable with the questioning and sceptical attitude of mind necessary for successful scientific enterprise. To be a good Christian, indeed to be a good citizen, required the unquestioning acceptance of authority. Even the arts saw their task merely as one of underpinning religious faith and endeavour. The great leap forward of the Renaissance was the discovery that free artistic enterprise and free scientific investigation, in fact the very freedom of questioning anything except faith itself, could be reconciled with religious practice and faith. The Renaissance, in a sense, discovered that scientific and artistic enterprise could be carried out in parallel with the exercise of religious faith and duties. What seemed an irreconcilable contradiction became reconciled. It was discovered that the mind was perfectly capable of working in separate compartments, the questioning compartment and the believing one, without being torn apart. Scientific and artistic enterprise was given freedom, and yet could serve the glory of God. Discovering laws of nature could be interpreted as discovering the will and the ways of God, rather than opposing church doctrine. The church did not accept these new departures without a fight and many men of the Renaissance found themselves in severe, often deadly, conflict with the church establishment. Indeed the church establishment found itself at odds not only with much science, but also with much theological thinking by the many reformers of the late Middle Ages and early years of the modern period. The reformers, such as Jan Hus (c.1370 – 1415), a forerunner of the reformation movement, and the main representatives of the movement, Martin Luther (1483 – 1546), John Calvin (1509 – 1564), and John Knox (c.1514 – 1572), to name but a few, were all critical of medieval Catholicism and demanded different interpretations and different practices of the Christian faith. They were all men demanding to be allowed to think for themselves, to go back to classical and biblical sources in order to re-interpret the Christian faith in their own way. The Reformation eventually led to more pluralistic interpretations of Christianity.

Desiderius Erasmus (1466 – 1536), known as Erasmus of Rotterdam, is another example of a deeply religious man who nevertheless demanded to be allowed to go back to classical texts and re-interpret them in his own way. He laid the foundations to the critical study of early Christian texts and to a new humanist, non-dogmatic, education. The educational reforms initiated, inter alia, by Erasmus, widened the syllabus to include classic languages and, more importantly, classic secular texts. His friend Thomas More (1477 –1535), equally embraced the ideal of classical learning, combined with free critical thought. As the famous author of Utopia, he may be regarded as one of the first political philosophers since the Greeks and the Romans, and particularly since Plato, who was active close on two thousand years earlier, around 400BC.

These founders of classical learning and critical thought, the humanists, were soon followed by the founders of modern scientific thinking, such as the philosopher Francis Bacon (1561 – 1626) and the father of experimental science, Galileo Galilei (1564 – 1642). The Renaissance was the watershed dividing the waning Middle Ages from the rising modern age. It marked the period of transition between blind obedience to dogma and to authority and the birth of the spirit of enquiry; between the closely confined thinking of the Middle Ages and the freely roaming thinking of the modern age. The juxtaposition of the intellectual atmosphere of the Middle Ages and that of the waxing modern age helps to create an impression of the limited capacity of medieval Europe for the creation of new thought, new knowledge, and new technology. The mind of medieval men and women was imprisoned in the straitjacket of an authoritarian doctrinaire Church and knew little intellectual activity outside the Church. Ordinary men and women were illiterate and utterly uneducated and followed the faith in blind belief, no doubt supplementing their meagre spiritual diet of bible readings and sermons with fairy tales and ghost stories told of a dark winter's night in a poorly lit, poorly heated, cramped home.

Only a tiny minority of people could read or write, and an even tinier minority had knowledge that extended beyond reading, writing, arithmetic and the bible. If ignorance is likened to darkness, then dark the age certainly was. For those of us who regard the period of the Enlightenment as one of shedding light, then by contrast the Middle Ages, with their strict adherence to church doctrine, and its medicine and other sciences based on Aristotelian principles rather than on experiment and logic, the period seems dark indeed. This is not to say that it was lacking in cultural achievement for a small class of educated men and women. Poetry thrived on a small scale, mathematics and astronomical observation made some progress, mainly owed to Arab men of science, architecture (especially cathedrals and castles) and the arts achieved great refinement and some unsurpassed pinnacles. The overall impression is, however, that humans lived in fear; fear of death and of hell, fear of sickness and starvation, fear of the authorities, and fear of war and plunder. If we define civilization as the liberation from fear, as well we may, we must regard the Middle Ages as rather uncivilized.

A gruesome example of the rein of bigotry and fear is the inquisition, first instituted in 1231 by Pope Gregory IX as a means of combating heresy, which by this time had become organised in large sects, such as the

Cathari and Waldenses, and was regarded as a threat to society. The rigidly hierarchical Roman Catholic Church felt seriously threatened by heretics, who disagreed on various points of doctrine and often expected Christian leaders to live a life of poverty. However, medieval inquisition at first imposed relatively mild punishments and the use of torture was not authorised till 1252. The worst cruelties and excesses of a fanatical faith, in the service of racial and religious intolerance, became institutionalised in the Spanish Inquisition, which began in 1478 and was not finally abolished till 1834, thus falling mostly without the medieval period. It is thought that under the rule of the first grand inquisitor in Spain, the notorious Torquemada, appointed in 1483, about 2,000 people were tortured and burnt at the stake and innumerable victims had their property confiscated.

No doubt the inquisition used ingenious technical means to inflict maximum pain on its victims. There can also be no doubt that here technology fulfilled the demands of the ruling elite. The demand for instruments of torture and their delivery by the artisans and inventors of the day is a classic case of market pull, or market-led innovation, albeit in the guise of a cruel caricature of the general principle. Not, I suspect, an example of market pull that even the most ardent believers in the benefits of market forces are proud of. It is undeniable, however, that technology, as ever, proved a faithful servant of the mighty and the powerful. In a very real sense, it was technology that made it possible for early hordes of primitive humans to become societies, and for wealth and elites to emerge. And ever since societies established hierarchies, technology has been a servant of the top echelons of society. Technologists have always produced what was demanded of them by those whose demands were backed by power and by money. It is only in more recent times that technology became more pro-active in the provision of goods and has become the prime mover of economic growth and the creation of wealth. Technology is now seen as the main provider of ever new means of making money and of ever new ways of creating a luxurious life. Unhappily these remarks are true only for the wealthy societies of this world and are almost wholly irrelevant for the world's majority of people, who live in dire poverty.

As in all preceding periods, wealth in medieval times consisted predominantly of land, though in the later Middle Ages large money fortunes were made from trade, and these laid the foundations to emerging capitalism. Most of the Middle Ages are characterised by the manorial system. Land was owned principally either by ecclesiastic or by secular large landowners, the lords of the manor. Bishops and abbots of monasteries could be lords of manors and hold large areas of land. On the worldly side the landowners were aristocrats and knights of various grades, many of them having risen to their position through military leadership. There were great variations in the size of the estates and in the division of the land between that farmed under the direct control, and to the sole benefit, of the lord, known as the demesne, and the land rented out in small parcels to tenant farmers, the villeins. The manor was a more or less direct successor to the Roman villa and, in Britain, passed from Roman into Anglo-Saxon and, after 1066, into Norman hands.

The tenant farmers were mostly serfs and thus were not only obliged to pay rent for their land and work on the demesne, but were severely restricted in their freedom. They were not allowed to move, marry, give away a daughter in marriage, buy property or do anything else of importance without permission from their lord. Their status was only slightly better than that of slaves and indeed many serfs were freed slaves. Others came into serfdom by seeking protection from a powerful lord against the vicissitudes of those violent and uncertain times. Some tenant farmers were freemen, and their number increased in the course of time as more serfs either purchased their freedom or obtained it as a reward for good services from their lord.

The tenant farmers lived in villages and farmed mostly open fields. That means that a large area of arable land was divided into smaller strips and each farmer was allocated a certain number of such strips, not necessarily forming a continuous area. The method of strip farming and the fact that the tenant farmers lived in clustered villages enabled them, indeed forced them, to cooperate in tilling the fields. The facility for cooperation made it possible to develop and use an advanced form of the Roman plough, the heavy plough known as the *carruca*.

As agriculture spread from the Mediterranean to the North, including Britain, the plough had to become able to tackle much heavier soils than the light soils prevalent in the Mediterranean region. The Roman plough generally had a wooden share, often tipped with iron. The later carruca had an iron share that could penetrate deeper into the soil. The carruca was usually equipped with a coulter knife, fixed in front of the share, that cut into the soil to create and direct the furrow. From about the tenth century, ploughs were equipped with a moulding board, mounted in parallel with the share, that turned the cut soil on its side. The carruca generally had wheels, thus greatly reducing the tractive effort needed to move the plough forward.

Another circumstance that necessitated the use of a heavy plough, and a strong team, was land clearance. In the early Middle Ages many forests were cut down, marshes drained, and other unproductive land taken into agricultural production. These new, previously untilled areas, constituted a challenge to the plough. The consequence of a heavy plough used on heavy soil was the need for stronger and better-harnessed draught animals. Oxen were employed in teams of up to five pairs and their harness consisted of a yoke across their withers, hitched from the shoulders and linked to the plough by traces tied to a swingletree. As a large team of oxen could not manoeuvre tight turns, it was necessary to plough large furlongs. All this added up to an operation run jointly by several families: a large team of oxen, an expensive plough, and the need for large fields. It was a technology compatible with farmers living in clustered villages and cooperating in tilling strips in large open fields.

In the course of the Middle Ages, beginning particularly from the late 14th century, the process of enclosure gathered pace. The number of farmers who held their land in severalty increased as the open fields gradually gave way to enclosed fields, worked by individual farmers. The process of enclosure continued for several hundred years and was not fully completed till the middle of the 19th century. As the areas of virgin land to come under the plough decreased and the size of fields diminished with enclosures, so the need for teams of oxen declined. In the meantime heavy powerful horses, capable of carrying fully armoured men, had been bred and pairs of such horses were gradually replacing teams of oxen. The harnessing of horses had improved and the horse offered a great advantage in speed over the ox.

As in any society and at any time, the social arrangements have to be suitable for the operation of the technology of the day. We can turn the argument round and say, with equal justification, that the technology employed must fit the social arrangements of the time. Technology and society form a system, and the system can function only if society and its technology are suited to each other to form a workable system, ideally an efficient one.

The large estates and the villages that housed the tenant farmers were the centres of social life during the early Middle Ages. Towns and cities were few and far between and trade was at a low ebb. The aristocracy held all the power in the land and its main interest was agriculture. Labour was cheap and, during most of the period, plentiful. Thus the incentives of the aristocracy for technological innovation or progress were minimal. In fact most of the lords were more interested in building castles and indulging in jousting and hunting then in productive investment. It needs to be said that castles were not pure luxury; they provided shelter in times of need and served as a base for the knightly warriors. The Normans built large numbers of castles that served as the military bases of their rule over England.

The lord of the manor derived his income from the sale of agricultural produce to the few towns and cities and to non-agricultural workers. In the fully-fledged feudal system, the lord of the manor also fulfilled many administrative and judicial duties of the state and was obliged to render military service as and when required by the king. The state functioned through the manors and the lords gave their services to the king in return for their land. Many of the lords of manors, particularly the owners of large estates, sub-let (sub-infeudated) parts of their estates to knights who, in return, were obliged to render military service for the manor as a whole. Alternatively, the lord could pay for knights to provide the service required of the manor and/or send his sons to do military service. Though the knights did not do full-time military service, they were nevertheless professional soldiers and formed the heavily armoured cavalry of the day. This was particularly true in the 12th century, though it continued till the knights became obsolete owing to the development of a superior infantry, armed with the longbow and the crossbow, in the 14th century. Military technology did not stand still during this period. The tenants were also obliged to render military service, whether as ancillaries to the cavalrymen or as infantrymen. There is a clear relationship between developments in military technology, i.e. the rise of heavily armoured cavalry, and the political system, which extended the more ancient manorial system into the arena of military service. The lord of the manor was converted into a knight and the knight into a landowner.

The military aspect of the feudal system, coupled with many religious, political, economic and social factors, is closely associated with the curious phenomenon of the crusades, which took place from the end of the eleventh to the end of the thirteenth century. They were a curious expression of the very essence of the medieval spirit, even though the causes of the crusades remain a moot point. A full understanding of a complex phenomenon such as the crusades may never be agreed upon among the specialists. It is clear, however, that religious fervour was one of the underlying causes. Perhaps this fervour was created by certain commercial or political interests, but it certainly provided very strong motivation for a large number of crusaders. To free the holy land from Islam and re-instate Christianity as the rightful faith was a fervent wish, and the belief that this was God's will was a fervent belief. The Middle Ages were a period when fervent beliefs were common – in fact religious fervour and strongly held beliefs are characteristic of the time. The second motivation (second merely in our listing, which is no reflection on the order of importance) was greed. The crusaders hoped to make fortunes by looting and in many cases these hopes were fulfilled. The spirit of adventure undoubtedly played its role with the romantic notion of reaching fabulous foreign lands in the service of God and King. Many of these dreams were left unfulfilled, as disease, hardship and enemy action killed large numbers of crusaders. The proportion of those who returned healthy and wealthy was far less that the proportion of those who never returned. As usual, the positive motivation – the pull – was supplemented by a push. Landless labourers and members of tenant farmer families were largely poverty stricken. They worked all hours of daylight and had very little to show for it. Poor housing, a poor diet, poor clothing and no money for any luxuries, even for the smallest expenditure beyond the barest necessities. Life was hard and monotonous; with practically nothing that might be called entertainment. Church and the bible and maybe some story telling were the only diversions. Apart from dreaming of adventure and heroic deeds, they dreamt of getting away from their arduous labour and from the dreariness of their lives. The desire to achieve something combined with the desire to escape from something. Strong motivation commonly consists of a pull and a push component: the wish to get somewhere and the will to get away from something.

In the upper echelons of the crusader movement somewhat different motivations were at work. On the one hand was the lust for power. Capturing foreign lands has always been a dream of the mighty. Enlarging the dominion, gaining power and wealth by expansion of the realm, have been goals of the mighty for as long as might itself has existed. Additionally, motivation to join a crusade was provided by the problem of the younger sons among the land-owning aristocratic families. The eldest inherited the estate, but the younger sons had to seek their fortunes elsewhere. The church was one outlet, armed service another. The image of the knight in armour riding against the enemies of Christ was a powerful image and a potent vision for many a younger son of wealthy families. The lure of the shining knight and the desire to get away from a dead-end situation at home provided the pull and the push needed for strong motivation.

Battles, exhausting marches and disease decimated the forces, but looting proved highly profitable. The hardships endured by the crusaders can only be imagined. The heat must have been unbearable, especially if armour of any description was worn. The havoc wreaked by vermin – lice, fleas, and other bugs – underneath the armour can barely be imagined. As opportunities to wash must have been rare, the stench of stale sweat defies imagination. The shining knight and the simple foot soldier were very likely soon reduced to smelly, itchy, bundles of misery. Add the hazards of infectious diseases carried by dirty water and vermin and caused by rotten or infected food, and one can imagine that enemy action was probably not the greatest hardship the crusaders endured.

The weaponry of the crusaders was essentially that developed by the Romans, though with some considerable modifications. The most substantial change was the greater emphasis on, and changed status of, the cavalry. The cavalry now consisted of armoured knights who had established for themselves a high social standing and had developed a whole mythology and code of behaviour known as chivalry. I do not suppose that the knights in the field bore much resemblance to the chivalrous knights familiar to us from medieval poetry. Development of heavy armour had continued apace, both with a view to technical improvement and with changes in fashion. The production of armour required high quality steel and great skills, and both of these had been developed in the early Middle Ages. Similarly, the sword had undergone much improvement and much change owing to fashion. The improvement was in the steel. A sword needs to be flexible rather than brittle and yet needs an edge that can be sharpened and will retain its sharpness. This posed quite a challenge to steel-makers and to sword-smiths alike. Fashion played quite a role in knightly equipment and indeed the role of fashion in technology in general should not be underestimated.

The politics of the crusades were extremely complex. The Byzantine Empire was surrounded by enemies on all sides and was glad to receive some help and relief in the shape of Christian soldiers. On the other hand, the schism between the Eastern and Western churches ran deep and thus the Byzantine Emperor had rather mixed feelings when hordes of Roman Christians turned up in his country. Some of the crusaders were disciplined soldiers, who pursued their goal of capturing the holy land, others were undisciplined hordes intent on plunder and were rather unwelcome in Constantinople. One of the crusades was diverted from its initial goal and turned on Constantinople instead, not exactly contributing to friendship between the Eastern and Western inheritors of the Roman Empire.

Many crusades disbanded, or were diverted, even before they reached the holy land and true success was achieved only for a relatively brief period by only a few of the many crusades. The very first crusade, known as the "Peoples Crusade", was soon annihilated by the Turks. The first properly equipped and organised crusade, known as the first crusade, started out in the summer of 1096, and consisted of four armies, with an estimated total of 200,000 warriors. The armies travelled by different routes, some travelling across the Mediterranean, some taking the land route over the Balkans. It is estimated that at one time there were about 4,000 mounted knights and 25,000 infantrymen in and around Constantinople, causing the Byzantine authorities more than a few headaches. By the time the crusaders reached Jerusalem in 1099, their number was reduced to about 12,500 men fit to fight.

Nevertheless, theirs was the major success of the crusades when, in 1099, they captured Jerusalem from the Muslims and founded the Christian Kingdom of Jerusalem. Jerusalem itself was re-captured by the Muslims under Saladin in 1187. The Christians retained a foothold in the holy land for another century, until their kingdom finally succumbed to Muslim attacks in 1291. The title of King of Jerusalem was kept alive on the island of Cyprus till the late 15th century.

From our point of view, the importance of the crusades lies in the demand they created for weapons, armour and transport, and the fact that European political and trading centres were re-established in the eastern Mediterranean. This in turn led to increased trade across the Mediterranean and to increased demand for shipping. Trade with the Byzantine Empire and the re-established European trade with the eastern Mediterranean brought many luxury goods onto European markets. Gold and silverware, fine cloths, religious relics, fine steel, and spices and aromatics were the main imports. Alas, the trade in slaves also flourished throughout the period. The main exports from Europe were timber, iron, copper and its alloys bronze and brass.

Disease was rife throughout the Middle Ages. The Black Death[1], that is estimated to have killed about one third of the population of Europe, arrived from the Middle East in 1347. It is said to have been brought to Europe by an early instance of biological warfare, when Kipchak warriors threw corpses of victims of the plague over the walls of a fortified Genoese trading post in the Crimea. Other common infections that occasionally reached epidemic proportions were leprosy, smallpox, tuberculosis, scabies, anthrax, trachoma, and cholera. It was a period when life expectancy was very low. Poor sanitary and living conditions allowed many infectious diseases to spread rapidly and ill-nourished people were easy prey for disease. Poor hygiene, poor living conditions, poor nutrition and medical ignorance caused infant mortality to be a constant scourge.

While the production of technological artefacts was based entirely on empirical knowledge – progress was achieved by trial and error and by accumulated experience – other aspects of life in the Middle Ages were dominated by dogmatic theories. The dominant theory was, of course, Christian theology. Even medicine was based largely on philosophy rather than on practical experience. The theories of Hippocrates (born c. 460 BC) and Galen (born c. 129 AD) dominated medicine into the late Middle Ages. There was a close correspondence between medical theory and Aristotelian theory. Thus health was thought to be achieved if the four humours in the body – corresponding to the four elements of Aristotle, i.e. fire, water, air and earth – were in equilibrium. The four humours were: hot and dry (corresponding to fire); cold and wet (corresponding to water); cold and dry (corresponding to earth); and hot and wet (corresponding to air).

The famous – or infamous – blood-letting (phlebotomy) must be seen as an attempt to restore equilibrium by removing surplus humours from the body. Blood-letting was used against acute illness and it is anybody's

[1] The term black death probably covered both bubonic and pneumonic plague.

guess how many deaths it caused. But it was also used as a precautionary measure. In some monasteries the monks underwent blood-letting up to once every six weeks. By the 16th century this was reduced to 4 or 5 times a year. Cauterisation, consisting of applying a hot iron to the body, mostly to the head, sometimes to wounds, was another technique designed to expel excessive humidity and coldness. However, apart from the techniques of phlebotomy and cauterisation, medicine did provide various herbal medicines against a variety of ailments. This aspect of medicine was certainly less painful and, hopefully, a little more successful.

The syllabus of medical faculties throughout Europe in the late Middle Ages consisted of a high proportion of philosophy and medical theory, coupled with medical practice. At Oxford University, as at most other universities, theology was the subject that attracted the most students. In the course of the 15th century, about 500 doctorates in theology were awarded, as compared to a mere 40 in medicine.

From 1169 clerics were officially banned from soiling their hands with blood, and thus had to withdraw from practising surgery. Medicine remained, to a considerable extent, in the hands of clerics and of monasteries. The edict was not always strictly adhered to, but nevertheless surgery became largely the domain of laymen, whether they be surgeons trained in universities or barber-surgeons, trained by apprenticeship, with fathers often passing on their skills and their businesses to their sons. Apart from cutting hair and shaving beards, barber surgeons were generally allowed to perform only the simplest of surgical operations, such as cupping (blood-letting), the application of leeches, the lancing of boils, the extraction of teeth, and similar.

Rudimentary anaesthesia was obtained by the use of the so-called soporific sponge. This was a sponge boiled in a mixture of opium and a variety of juices, including hemlock. The sponge was placed under the nose of the patient and induced a deep sleep. To wake the patient, a similar sponge, soaked in vinegar or similar, was placed under the patient's nose.

Anatomy was a much-neglected subject, as the dismembering of the human body was forbidden by the church. Because of the objections of the church, the proper study of anatomy was very slow to develop and most medics and even surgeons had to make do with learning from books based on Greek and Arab sources, or animal cadavers. In the later Middle Ages a limited number of bodies of executed criminals became available for the instruction of students of medicine and surgery, but even then the students had to be content with watching rather than doing.

Surgeons were quite skilled in setting bones, but also tackled some quite complex operations, such as the removal of gallstones, which seems improbable when one considers the absence of anaesthetics and hygiene. Surgeons were in charge of all injuries, which were plentiful. Brawls were common, and wars were a permanent feature. Thus surgeons had to dress wounds, remove lodged arrowheads or bolts from crossbows. Occasionally they even had to mend broken skulls. Presumably they were often called upon to amputate limbs shattered beyond repair or afflicted with gangrene. The highest-ranking surgeons accompanied kings and knights into war and enjoyed a high status and high rewards.

Apothecaries formed the third strand of medical provision. Much pharmaceutical knowledge was based on the writings of classical Greek scholars, greatly expanded by Arab contributions to the subject. Virtually all remedies were herbal, at least in the sense that they were of vegetable origin. Simple people had to make do with homegrown herbs and spices, whereas the wealthier classes, including the monastic orders and the church hierarchy, used many imported remedies that were extremely costly. Then as now, people have been willing to spend up to their ultimate financial capabilities if they believe that the expenditure will benefit their health. Health has always been a profitable business.

One of the imported medicines was sugar, much favoured as a syrup for chest complaints because of its warm and moist nature, but also used to sweeten bitter medicines. Sugar had to be imported, as the only known raw material for its manufacture was sugar cane, a plant that grows only in tropical or sub-tropical regions. The manufacture of sugar from European sugar beet was not invented till the middle of the 18th century, and did not reach industrial production till the 19th century, when the nascent sugar industry in France obtained much support from Napoleon in an effort to overcome the effects of the continental blockade. This is a nice example of direct government intervention in the course of technology. Technology policy – the support of a particular technology – became part and parcel of economic policy at a time of war.

Much medical and pharmaceutical knowledge came to Europe from the Arabs, where these sciences first flourished. The first medical school in Europe was established in Salerno, followed by Montpellier, Paris, Bologna and Padua. The rest of Europe followed in due course. The school of Salerno was also the first to obtain an official seal of approval from the Holy Roman Emperor Frederick II, who decreed that no-one should practise medicine without passing an examination by the Salerno school, thus obtaining an official seal of competence. This may have been a simple case of patronage, but more likely it was an early example of state control over a professional activity that had plenty of potential for harm. Whether in those days a licensed medical practitioner was less damaging to health than an unlicensed one may be a moot point, but it is almost certainly an early instance of the state taking upon itself the regulation of a profession for reasons of safety. This role of the state has been retained to this day, and indeed has been greatly strengthened and expanded. All professions that are potentially dangerous to the health or the safety of the public, including, inter alia, medicine and civil engineering, are now strictly controlled in some way by the state.

The immediate post-Roman period in Europe was characterised by incessant conquests by so-called barbarian tribes, causing havoc in the cities. Feudalism became established in Europe in the 9th century as a result of a general shift of civilization away from towns and cities into the rural domain. City populations declined, and merchants almost disappeared. Long distance trade declined in consequence of the loss of the Mediterranean to European traders, as the Middle East became dominated by the Muslims. The decline in trade over middle distances may be related to a general decline in commercial life as safety in towns and on the roads declined, and neighbouring provinces in many parts of Europe constantly fought each other. There was no powerful central authority and all the Romans had built up was in decline. The roads – the famous Roman roads – virtually disappeared for lack of maintenance. On the other hand, robbers and minor local authorities demanded extortionate road-tolls and transport and travel declined. Power and wealth became concentrated in the hands of feudal landowners and their vassals. Virtually all manufacture took place on the large estates that had to be self-sufficient in fulfilling the modest demands of their inhabitants. Markets became small, insignificant and purely local. The only commodities that were traded over greater distances were salt for the general population and some spices and precious materials for the few rich who could afford them.

The feudal estates produced all agricultural produce available in the region, as well as all other essential products, such as agricultural implements, textiles, furniture, domestic equipment, and so forth. Goods were sold in small-scale local markets and their producers were strictly tied to the estates. Thus technology catered purely for the basic needs of a rather poor and simple rural population, consisting largely of serfs and some free tenant farmers. No doubt the feudal landlords and their managers, as well as a few professional people, demanded a little more material comfort and even some luxuries, and these were supplied by a few remnants of a wider ranging trade. In local emergencies, such as droughts and consequent famines, produce was procured over longer distances, but these circumstances were the exception rather than the rule.

The estate-based crafts might occasionally be called upon to produce some higher-class items for the local elite, but basically they catered for essential needs and were not subject to either competition or the influx of foreign ideas. Thus the early part of the Middle Ages remained at a technical stage reached by the Romans or, if anything, declined from that achievement because of the decline of urban demand and demand by a highly organised large military machine. Only a few areas of technology showed progress, especially personal armour, swords, and some agricultural techniques.

The period of instability following the collapse of the Roman Empire began to be overcome by the end of the 10th century and a period of population growth began that lasted well into the 13th century. The worst of the aftermath of the collapse of the Roman Empire was over and the political and social situation became sufficiently calm to allow sufficient food to be grown and distributed, so that the birth rate exceeded the death rate. Because landlords, mostly large ones, owned all the fertile land and, apart from owning and tilling land, there were few other means of making a good living and becoming prosperous, the growing population was forced to search for new economic opportunities. One way of creating opportunities was to bring more land into cultivation by draining marshes or by felling trees. Indeed some men escaped from serfdom and cleared and drained previously unused land and settled on it as free peasants. The Cistercian monks often founded monasteries in the wilderness and brought new land under cultivation, creating new opportunities for the monks and for their lay tenants. Many new villages were founded on land owned by a landlord, and the new villagers became tenant farmers rather than serfs. As commercial life revived, the landowners became interested in monetary income from rents and feudalism slowly declined.

As the land offered few opportunities for free men, there was a tendency for people to move into towns and engage either in commerce or in crafts. The 12th century saw a rise in commercial activities and a considerable growth in the merchant class. Many fortunes were made by trade. Some serfs escaped from the estates and found refuge, opportunities, and eventual freedom in the towns. The towns offered many advantages to their citizens. They had freedoms that peasants and country dwellers could only dream of. Townspeople could travel freely, make fortunes and keep them in the family, and they had opportunities to engage in trade or in a craft.

Economic activity, including the manufacture of technical artefacts, gradually shifted from the estates to the towns. In the towns technology was in the hands of artisans and was highly organised. Competition in the present sense had not entered into the thinking of the time. The belief was that the role of the guilds – the associations of artisans – was control over the quality of entrants into the professions, control over the quality of the products, and the safeguarding of fair play between all members of the group and with the consumer. The concept of fair play included the notion of a fair price for a fair product; thus price cutting and cutting of corners were equally frowned upon. Specialisation and competence were the dominant ideas, not the search for novelty. The number of trades had become quite large and the products had reached a high standard of quality, but there was little incentive for technological innovation. Markets were not saturated; the only brake on sales of technological products was lack of money, not lack of interest, need, or demand on the part of customers. The organisation of manufacture was in very small units. A master, a few journeymen, i.e. fully trained workers who had not achieved the status of master, and a few apprentices made up the full complement. To rise from journeyman to master required not only experience – often acquired by travelling from master to master – but also needed capital for the acquisition of premises, tools and materials. Apart from all that, numbers were restricted by the guilds in order to avoid a ruinous over-supply of craftsmen. One way of obtaining the status of master was, of course, for a journeyman to marry a master's daughter and take over the business. The sale of craft products to the public was direct – no large showrooms, very few, if any, middlemen. Local people bought directly from the local craftsmen, only more distant customers were served by travelling traders.

In the later Middle Ages and with a revival in trade, some goods were produced in large quantities, particularly cloth of various kinds. As the scale of operations increased and more capital was needed to invest in raw materials and equipment, manufacture became organised around merchants who were able to raise the capital. The actual production was still carried out in small units specialised in various aspects of production: carding, spinning, weaving, dyeing, and finishing. Thus trades such as fullers became established. The merchant supplied the raw materials and provided what we now would call logistics. They organised the progression of work, the transport of intermediate products, and the sale of the finished products. Various regions and towns became specialised in various products. Obviously the regional availability of raw materials and tools and the possibilities of cooperation between different producers provided many advantages that outweighed the disadvantages of local competition.

The power in the land shifted to some extent from the feudal landlords to the towns and to the artisan and merchant classes. This process continued until, by the late Middle Ages, feudalism had given way to rising capitalism.

As demand for goods increased in the later Middle Ages, some incremental – as opposed to radical – technological progress occurred. Looms grew bigger and the length of cloth woven increased. The whole of Flanders became a country of weavers and fullers. The centres of production of woollen cloth were towns such as Ghent, Bruges, Ypres, Lille, and Arras. We can see that some of the specialisations of those remote days have outlasted the centuries, and a town such as Lille is still a centre of textile production. When the industry had outgrown the local production of wool, wool was imported from England. Woollen cloth from Flanders or Brabant was superior to other regions in finish, softness, and colours. Because transport was expensive, only the highest quality goods were transported over longer distances; cheaper products were only sold locally.

Other towns and regions specialised in different products. The Meuse valley became a centre of copper working in the 11th century. The town of Tournai (in modern Belgium) became a centre of stone working and provided baptismal fonts far and wide. Lucca in Italy became a centre for silk weaving, while Milan and the Lombardy towns produced mainly cheap fustian.

The degree of concentration of certain trades in some towns is remarkable. Thus Ghent had 4,000 weavers and 1,200 fullers out of a total population of about 50,000. Similarly, in Ypres nearly 52% of the population were engaged in the woollen trades.

The advantages of cooperation apply to this day. We still see concentrations of certain industries in certain places. Perhaps the best-known example is the so-called Silicon Valley in California, which is one of the main centres of production of modern electronics based on silicon chips. Although the very first firms established in the region may have chosen it for a variety of reasons, the founders of the later firms producing chips deliberately chose the proximity of similar manufacturers because of the possibility of sharing information, sharing suppliers and, occasionally, sharing sales contracts[2].

We also see remnants of former concentrations of tradesmen in street names such as Carpenter Street, Cloth Street, Fuller Street, Goldsmith Road, and so forth; but also in the name of certain neighbourhoods, such as jewellery quarter or in the continuing tradition of certain roads, such as Savile Row in London continuing the tradition of fine tailoring.

The concentration of certain trades in certain areas held advantages for both the trades-people and their customers. If you needed a pair of shoes, you simply went to Shoemaker Lane and were able to compare the shoes and the prices of different shoemakers. The craftsmen profited by being able to help each other out, and benefited from the availability of the supplies they needed and by exchange of experiences. Helping out was sometimes of crucial significance, particularly in the case of armourers, as the demand for their wares could arise very suddenly and urgently and far exceed the capacity of any one master. The reverse side of this coin was, of course, that each craftsman was able to spy on and supervise all the others, so as to eliminate malpractices.

Craftsmen not only manufactured items of ordinary consumption, they also were in charge of large enterprises, such as major buildings. The person in charge of stone buildings was a stonemason; whereas timber buildings were built by carpenters. In a big enterprise, the main contractor employed numerous sub-contractors of varying skills. Stonemasons were not all equally well qualified. Some were entrusted only with the rougher work, and only the best undertook the shaping of the elaborate stone ornaments and sculptures that we admire in many a medieval cathedral. Even the largest buildings were designed and constructed by artisans, there was no such thing as an architect or a civil engineer. Designs were based on experience and on talent, not on theory.

The 12th and 13th centuries saw a considerable increase in the circulation of money. With the rising importance of trade and of towns, and an increased availability of a variety of goods, people needed money to take advantage of the new opportunities. The land-owning aristocracy discovered the need for money to buy luxuries, where previously they were content to make do with the supplies produced on their estates. The quality and quantity of food and of dress improved, household furniture became more elaborate, and the taste for luxuries increased. The feudal landlord was happy to sell freedom to his serfs and to derive income from his tenant farmers. Agricultural production became more specialised and tenant farmers were interested in purchasing more land. Even townspeople with surplus cash invested in land – the ownership of land was suddenly seen as a profitable commercial transaction. Whereas in feudal times the ownership of land implied power over a certain region, these implications gradually ceased in the 12th and 13th centuries and land became a commodity and a source of income without too much social meaning.

The early long distance merchant travelled with his wares, often in armed caravans. As trade between countries to the North of the Alps and Italy developed, improvements were made to the land routes across the Alps. For example, Europe's (and probably the world's) first suspension bridge was built across the St. Gotthard gorge in Switzerland at the beginning of the 12th century. In the second half of the 13th century, transport became divorced from trade. The merchant became sedentary and the transport of goods was undertaken by specialists. This new division of labour was accompanied by several developments. First, transport itself developed. Not in one fell swoop with a major innovation, but gradually with improvements in roads, improvements in road ve-

---

2 See e.g. Braun and Macdonald, Revolution in Miniature, (1982)

hicles and, probably, better breeding of suitable horses. Shipping improved quite radically with increasing size of ships, to about 200 to 600 tonnes, and an improved design of rudder. More importantly, the introduction of the mariner's compass freed ships from the need to follow coastlines and enabled them to take the shortest route to their port of destination. During the 14th century the compass became universal in the Mediterranean, and during the following century it came into general use in the North and Baltic Seas. The ports were equipped with wooden quays and with cranes.

The second consequence of the sedentary merchant was the need for organised financial transactions. The merchant needed credit to finance large quantities of goods in transit and, a little later, he needed possibilities of paying for goods in distant lands without the need to transport large quantities of cash. Credit in medieval times was problematic because of the church's opposition to usury. Thus usury, although practised on a large scale from about the middle of the 12th century, had to be disguised and atoned. The church made sure that merchants who made profits out of loans suffered from a bad conscience and atoned for their sins by making large gifts and/or legacies to the church. There were other clever means of avoiding the charge of usury. It was possible, for example, to make a loan in one currency and obtain repayment in a different currency. A change in the exchange rate of the two currencies during the lifetime of the loan could provide the lender with a profit justified by the risk of currency fluctuations, rather than based on payment of interest.

Spices were one of the first objects of international trade. They generally came from India and could either take the land route to a Mediterranean port or could travel by sea to reach a Red Sea port, before being transported by land across Egypt to the Mediterranean coast for shipment to Europe. In 1498 Vasco da Gama found the sea route from Portugal to India by sailing round the African continent, thus justifying the name Cape of Good Hope for the southern tip of Africa. This route was developed by the Portuguese during the sixteenth century and offered considerable competition to the mainly Italian merchants using the old routes across the Mediterranean. Pepper, cinnamon, cloves, nutmeg and sugar-cane entered the diet of the wealthier Europeans. From the beginning of the 13th century imports from the East into Europe included rice, oranges, apricots, figs, raisins, perfumes, medicaments, dyestuffs such as Brazil wood from India, cotton, silk, damask from Damascus, baldachins from Baghdad, muslins from Mosul, gauzes from Gaza. This trade was balanced by exports from Europe of timber, arms, slaves, and woollen goods, particularly fustians from Italy and cloth from Flanders and Northern France.

By the 13th century there was an established market for luxury goods, such as exotic fruits, perfumes and luxury cloth. The people who were able to consume such goods belonged either to the landowning or, increasingly, to the merchant classes. The families that had made fortunes from trade soon formed a separate upper class. They built better houses for themselves, consumed luxury goods, including, no doubt, gold and silverware, and they formed into groupings that exercised considerable power in society. The gilds or hanses of the wealthy patriciate were born. Much of the governance of towns and cities became concentrated in the hands of this new urban upper class.

We mention medieval cities and imagine them as large entities. This was not the case. The only cities approaching populations close to 100,000 in the 14th and 15th centuries were Venice, Florence, Milan and Genoa. Northern European cities in the middle of the 15th century were much smaller:

Basel	about 8,000 inhabitants
Brussels	about 40,000
Frankfort	about 9,000
Nuremberg	about 20,000
Strasburg	about 26,000

The urban populations became stratified: wealthy merchants as the upper class, artisans and small traders as the middle class, and all the rest as the lower class. Rare and luxury items were produced only in the larger towns; smaller towns produced only items of everyday consumption for the town and its rural surroundings. The rural surroundings, in their turn, produced all the food the town required. The common trades were bakers, butchers, tailors, blacksmiths, joiners, potters, pewterers, and many more. The relationship between the gilds and the authorities varied and was not always smooth. Eventually, in the course of the 14th century, the gilds reached a position where they participated in local government and had considerable autonomy in regulating their own affairs. The gilds regulated entry into the professions and the training given to apprentices. They also regulated working hours, wages, and prices. It may be said that the gilds stifled competition and prevented innovation; on the other hand, they guaranteed the quality of workmanship and protected apprentices and journeymen from exploitation. The free artisans provided the goods required in the nearer or wider locality, whereas goods for long-range trade were often produced by craftsmen employed by the merchants. As capitalism became more mercantilistic and more aggressive, some conflicts, including strikes, between workers and their masters became inevitable.

For those modern contemporaries who believe unconditionally in the blessings of technical and scientific progress and in the power of competition to spur people to greater achievement and greater happiness, the Middle Ages are abhorrent on account of lack of competition and technological stagnation. The gilds would be anathema to contemporary neo-liberals. Their purpose was to protect their members from too much competitive pressure and to protect the public from both shoddy workmanship and excessive prices. The gilds protected the good name of the craftsmen and in this way protected the public from unreliable suppliers of goods. Considering that the *caveat emptor* (buyer beware) principle now rules almost supreme (though mitigated by some statutory regulation and by guarantees), and the buyer is often unable to beware sufficiently and falls victim to all kinds of malpractices, the security offered by the organised medieval craftsmen does not seem like something worthless. Buying goods in the certain knowledge of receiving value for money is worth a great deal. One can but wish that modern merchant bankers had been organized in guilds before causing so much trouble in the credit crisis.

The 14th century generally saw a halt to economic progress and a stop to population growth. Disasters struck Europe in the form of famine and the notorious Black Death that killed about a third of the population of Europe between 1346 and 1369. Armed struggles between neighbouring states added to the calamities. Italian cities fought each other and Germany sank into anarchy. The Hundred Years War (1338 – 1453) between England and France very nearly ruined both. As times became harder, towns made it more difficult for "foreigners" to become citizens. They also made it harder, indeed illegal, to carry out trades outside the city walls. Only spinning was allowed in the countryside, the rest of the textile manufacturing operations had to be carried out within the cities. It all sounds rather familiar: a mixture of protectionism and immigration controls as a response to a harsher economic climate. It did not always work and, in any case, it did not at this stage apply to large cities and to states. By the beginning of the 15th century, competition from English textile manufacture became very strong and Flanders lost its predominant position.

Despite all these problems, the 14th and 15th centuries saw the growth of large commercial companies that engaged in long-distance trade and in banking. The first such companies started in Italy, but soon found competitors north of the Alps. Probably the first major modern bank, Casa di S. Giorgio, was founded in Genoa in 1407. This was also the period when protectionism gradually shifted from the towns to larger entities. Economic power shifted from municipalities to states. In England, Henry VII (1485 – 1509) pursued vigorously mercantilistic and protectionist policies started by his predecessors. The importance of industry grew rapidly and became a prime object of state economic policy. The state attempted to protect English production from foreign competition, to protect and enhance English shipping and English exports, and to reduce imports. Whereas previously Flanders used some English wool to supplement its own sources for its textile industry, by now English cloth producers consumed all the homegrown wool and exported cloth.

We have excellent contemporary descriptions of some of the technologies current in the middle of the sixteenth century and early seventeenth century AD. Although strictly speaking this period belongs to the Modern Age, rather than the Middle Ages, I shall take the liberty of describing these technologies as medieval because we may safely assume that they were current in the late Middle Ages. The pace of technological change was not as rapid as in later centuries and what was described in a standard text in 1550, or even 1620, may safely be considered late medieval technology.

Our knowledge of the state of metallurgy and mining of this period is based mainly on the famous text by Georgius Agricola, *De Re Metallica*, first published in 1556. The German Georg Bauer was a man of his time and used the Latin translation of his name as a nom de plume. His book is a product of his time in that it contains detailed technical information alongside expressions of piety and general wholesome advice. When he describes the desirable character of an owner of a mine, he mentions piety as the first requirement and prudence, in the sense of spreading risks between several mines rather than owning a single one, as the second most important characteristic. On the other hand, Agricola does not believe in the power of divining rods to find ores, but instead puts his faith in the power of geological knowledge and observation. He sings the praises of metals and is convinced that their use is a blessing for humanity. By way of good advice, Agricola suggests that "It is necessary that the assayer who is testing ore or metals should be prepared and instructed in all things necessary in assaying, and that he should close the doors of the room in which the assay furnace stands, lest anyone coming at an inopportune moment might disturb his thoughts when they are intent on the work."[3]

He provides a detailed description of underground mines, presumably for various metallic ores. Underground mines are constructed by drilling vertical shafts connected to horizontal tunnels. The miners reach the tunnels on ladders and the ore is lifted in buckets by winches. The winches are mostly operated by hand, though some are powered by horses moving in a circle or working a treadmill. Water has to be pumped out of the mine and a variety of pumps may be used for the purpose. Some pumps consist of a chain of buckets; others use the principle of a piston operating in a cylinder. The cylinder might be constructed by boring out a log with an auger drill. The number of different designs of pumps used from antiquity and right into the modern age is amazing. From the simple single bucket pulled on a rope, with or without a pulley, or lifted by a lever construction, right up to elaborate piston pumps. An entirely different design principle was the Archimedes screw, where water is transported and lifted by a metallic or wooden rotating spiral. Some piston pumps were operated by a screw and toothed rod, which turned rotary motion into the to-and-fro motion of the piston.

There was an acute awareness of the dangers of gas formation in mines and the need to ventilate the tunnels. Various devices were used to suck out, or drive out, the pestilential air. Ventilation could be achieved either by diverting wind into the shaft, or by using fans that could be driven by a water wheel, or by bellows. The latter could also be driven by water or by horses or other animals working a treadmill. Miners were aware of the disease afflicting their lungs, quite apart from the dangers they faced through collapses of tunnels or explosions of methane gas.

The book contains much detailed information on metal smelting. The metallic ores are first broken up, sometimes using mechanical hammers driven by water or animal power, and then ground and heated (roasted) below the melting point and calcinated. After all this preparatory work, the actual smelting process is carried out. We describe here the smelting of iron. A crucible is placed in the hearth inside a closed furnace. The hearth is made of powdered charcoal mixed with powdered clay. The furnace is filled with the prepared ore, charcoal and unslaked lime. Air is blown into the furnace by bellows and the slag is allowed to run off. After about 8 to 10 hours the molten iron is poured out and when it had cooled a little, it is compacted with a hammer to drive out remaining slag. A higher class of iron, we might term it steel, is made by melting iron over a long period in a crucible with the addition of charcoal. It is then again treated by hammering it and eventually tempered in cold water.

By this time a whole array of flourmills had been invented. Some were driven by men, some by animals, some by water and others by wind. They came in all shapes and sizes; some were even portable and could be operated by a single man. Power-driven saws for cutting timber and even some for cutting stone were available. Any number of different designs of lifting gear: cranes, multiple pulleys and various combinations thereof for loading and unloading ships or carts, were in use.

Much ingenuity was spent on siege machines. Some were designed to bridge ditches; others provided an artificial steep slope to enable troops to move up the incline to the top of a fortification. Other machines were designed to ram gates or walls, and yet others to sling missiles over great distances into fortified towns or castles. It was a time when fortifications of all kinds were built and ever further refined. Virtually all towns had walls, and large numbers of castles were built. As fortifications became more refined, so siege machines were improved in the usual manner of offensive and defensive weapons developing in parallel and keeping some sort

[3] Agricola, Georgius (1912). De Re Metallica, translated from the first Latin edition of 1556 by Herbert Clark Hoover and Lou Henry Hoover. London: The Mining Magazine, page 223

of balance. The difference between then and now – apart from the complexity of the weapons themselves – was that defensive and offensive weapons systems could be clearly distinguished, at least as far as fortifications were concerned. A siege machine is clearly offensive, a fortress clearly defensive. Similarly, a shield is defensive and a sword offensive – though the distinction breaks down here because a sword could be used for defence as much as for offence.

Some devices of the time must be regarded as toys rather than useful machines, though the production of toys and ornaments has always been one of the lesser goals of technology. Ornamental water fountains of some complexity were built as the art of the stonemason combined with the arts of producing water pipes and pumps. A machine was invented for holding several books open for the reader, who could rotate the machine to access different books.[4] We do not know whether anybody ever used this weird contraption.

The Romans had a well-defined system of measurements. Medieval Europe more or less continued with the Roman system, but it was less uniform because there was no overall authority as there had been in Roman times. The main units of measurement were the ell, a unit of length used extensively for measuring cloth. The great trade fairs of the 12th and 13th century kept their own standard ell. The ell of Champagne was based on an iron standard kept by the Keeper of the Fair. It measured about 76cm (30inches) and was accepted by many important centres of the woollen trade, such as Ypres, Ghent, and Arras. The ell in other towns was slightly different. The Roman unit of weight, the libra, survived into the Middle Ages, as did the mile, though the pace, on which the mile was based, disappeared and gave way to feet and yards. Liquids were generally measured by the pint, which was roughly equal to the modern litre[5].

The metal industries developed gradually through many parts of Europe. One aspect of this development was increased use of waterpower for driving bellows, hammers, and stamping mills. The scale of production of iron increased, the knowledge and skill of metal workers improved both with experience and with the slowly growing availability of specialised literature. The temperature and size of smelting furnaces increased. Cast iron became an important part of iron production and the smelting furnace developed into the blast furnace. The production of other metals also increased and improved. For example, tin mining in Cornwall rose steadily from the 10th century. Coal mining also increased. Iron smelting still required charcoal, which became increasingly scarce with large-scale deforestation and the demands on land made by agriculture, especially in the Mediterranean region. Coal was used for preliminary operations and for lime burning and, of course, in forges. Coal mining centred particularly on Liege from the end of the 12th century, and Newcastle from the 13th century. Coal was shipped from Newcastle to London (so-called sea-coal) and was used there for domestic heating, with consequential problems with, and complaints about, smoke and smell.

The so-called Stückofen (lump furnace) provides a good example of production gradually improving with experience. This type of furnace produced a solid lump (bloom) of metal that was extracted from the top of the shaft at the end of the smelting operation. The weight of the bloom produced in a single smelting operation increased from about 10 kg in the earliest days of the Iron Age to 370 kg in 1430, to 400 kg by 1470 and to 500 to 600 kg by about 1600. By then, however, the method of producing iron in bloomeries was obsolescent. The Stückofen had been largely replaced by the 15th century development of the blast furnace, a shaft furnace that used somewhat more charcoal, but allowed the molten iron to run off through a tap hole. This iron could be used directly as cast iron or could be turned into malleable wrought iron by treating it in a separate hearth, the so-called finery hearth. The iron from the blast furnace was often initially cast into shapes reminiscent of suckling pigs and became known as pig iron. The initial impetus for the production of cast iron came from its use for the casting of cannon. Cast iron cannon was much stronger than cast bronze cannon and much easier to manufacture than forged iron cannon.

Blacksmiths became highly specialised. Some made wire, others made needles; some made swords, others made scythes; some made anchors, others made horseshoes. Up to the end of the 10th century wire was made by forging. The draw-plate for wire, which allowed wire to be produced by drawing hot iron through a suitable hole, was invented at this time and drawn wire soon replaced forged wire. Early iron needles had a hook,

4 Ramelli, Agostino. (1976). Various and Ingenious Machines.

5 One litre is approximately two modern British pints; now largely obsolete.

rather than an eye, and were produced by hand by members of the guild of needle-smiths in Nuremberg in about 1370. The first needles with an eye were made in the Netherlands in the 15th century. Swords were made in many centres, eminently in Milan, Brescia and Passau. The craft came to Solingen as a result of the Italian campaigns of emperor Frederick I Barbarossa (1152 – 1190). Scythes were made primarily in Styria, the birthplace of iron manufacture and a rich source of iron ore.

By far the greatest numbers of medieval people were engaged in agricultural pursuits and this division of occupations hardly changed up to the time of the Industrial Revolution. Yields were small, productivity was low, agriculture was extremely labour intensive. It thus required the largest part of the working population to work the land in order to feed the population as a whole. We have seen that Roman agricultural implements had made considerable progress in comparison to earlier times. The Middle Ages saw some further advances, though no radical change. The process of land reclamation continued to some extent into the 13th and even the 14th centuries.

It is interesting to compare yields of agricultural crops during the Middle Ages with modern yields by way of illustration of gradual technological advance. The average yield of wheat on modern fields is in the region of 40 to 50 bushels of wheat per acre[6]. When a field was left experimentally without manure from 1843 to 1967, the yield declined to about 12 bushels per acre. This is still twice or three times higher than the average yield of a well managed and manured field in the 13th century. The difference lies in better methods of cultivation and in plant breeding having provided better varieties of wheat, rather than in any dramatic change in technology. The combine harvester and the tractor have drastically reduced the need for manpower and for hard physical work, and have increased the energy consumption of farms, but have not changed the yields all that much.

One of the problems of English agriculture was the balance between pasture and arable land. In the days before artificial fertilizers, the fertility of the soil had to be maintained by rotation between different crops and fallow, and by animal manure. Hence arable land could not be expanded indefinitely at the expense of pasture, because a sufficicient number of animals were needed to maintain the fertility of the arable land. At times, the balance between pasture and arable land became precarious and the price of pasture rose to levels above those for arable land. In addition, pasture was needed for the production of wool and wool was one of the chief exports of England. In earlier times, much of the raw wool was exported to Flanders, where Europe's foremost woollen industry was established. In the course of the 14th century, however, the export price of English wool rose considerably because of large profit margins of the merchants and high taxation imposed by the crown. The high price of the raw material forced the Flemish industry to concentrate on high quality products, thus limiting its markets to the luxury end. The high price of the raw material also became a contributing factor to reducing the wages of Flemish workers. Social unrest resulted, which led to an urban revolution in 1320 and to the emigration of many Flemish textile workers to the Brabant, to North Holland and to England. To some extent because of the influx of skilled Flemish workers, but also because of the realisation that more profit could be made and more employment provided by substituting the export of cloth for the export of wool, English textile production rose sharply in the course of the 14th century. There were some half-hearted attempts by the English crown to force this process by forbidding the export of raw wool, but the prohibitions did not last long and had no substantial effect. Even so, the attempt to direct the economy away from the export of raw material and toward greater industrial production, half-hearted as it may have been, must be seen as a rudimentary effort by the state to introduce economic and technology policies.

One of the reasons why the feudal system began to give way to a more market oriented society, though perhaps not the main reason, was the obsolescence of the armoured knight. The armoured knight was too heavy and too clumsy to be very mobile and became helpless if unseated. More lightly armoured cavalry on faster mounts proved, in the end, more effective. The death knell to the knight was sounded by two technological innovations: the crossbow and the longbow. The crossbow was very effective in disposing of even heavily armoured cavalry and infantrymen could easily learn to use it. The famous English longbow was an even more effective weapon, but the acquisition of the required skills to use it effectively was a lengthy process. At the bat-

---

[6] This corresponds roughly to four cubic metres per hectare.

tle of Agincourt (1415), the more numerous French army proved no match to the well-trained English longbowmen.

The more important factor that spelt the end of feudalism was the development of trade and the rise of towns. Between the 11th and the early 13th century numerous towns were founded, largely as trading and market centres, but also as centres for pre-industrial craft production. The citizens of the towns, the burgesses, were freemen and the towns soon developed their own systems of self-government. If a serf managed to escape and enter into a town, which was not easy, he eventually acquired his freedom. Much technology developed in the towns. Town walls were built, and the crafts developed. In England, the walls were built not so much as protection against armed attack by armies, but mainly against the infiltration of undesirable elements into the town. The towns jealously guarded their right to choose who shall and who shall not be allowed to dwell in them, or even spend the night in them. In many other countries warfare between neighbouring rulers was common and city walls did serve to protect the towns against major armed conflict. In Germany and Italy major cities became the centres of states and the small city-states remained, in one form or another, right into the 19th century, when both Italy and Germany were unified.

The main function of the towns was, of course, their role as trading and market centres. The towns had to be supplied by the surrounding countryside with food and other necessities, such as firewood, and these goods were sold in markets. Only salt and a few spices were brought from further afield. As more burgesses became affluent, more imported goods appeared in the towns. Wine, more exotic spices, medicines, ornaments, fine cloths, and so forth. Some traders who were in business selling goods that originated in distant places, or even overseas, settled in the towns. The towns provided not only safety, but also the infrastructure needed by traders. They often needed mutual support, they needed rudimentary financial services, they needed transport services, legal services, schools for their children, housing and business premises. With expanding trade and increased prosperity, the towns expanded to provide the services required. The towns also became centres of craft production, with all kinds of artisans settling in them to produce and sell their wares.

The Middle Ages were a period of strict hierarchical order and strict adherence to the teachings of the Church. Science made virtually no progress, as experiment and observation were regarded as superfluous. The doctrine of the Church and abstract theories, based on church doctrine, sufficed as explanations for all phenomena requiring an explanation. Thus medicine was based on abstract theory and it was firmly believed that both sickness and health could only come from God. Hence medical practice was always reinforced with prayer and the intervention of various saints was regarded as essential in warding off sickness and restoring health. Astronomy was based on Aristotelian principles that had the approval of the Church.

On the other hand, much activity was based on pure empiricism. Surgery, as opposed to medicine, was based on practical experience, with theory counting for little. All the trades, the manufacture of textiles, building, furniture making, and the rest were based on practical knowledge passed from generation to generation without too much development, experimentation, or innovation. The emphasis was on good sound practice, not on novelty. This is not to say that there was no technological development, but development was slow and was not deliberately sought as it is today. There was no premium on novelty. Perhaps one remarkable medieval invention ought to be singled out for mention: spectacles. The first known spectacles were invented in Italy in the 13th century. The Bishop of Exeter acquired a pair in 1326. The spectacles consisted of bone eyepieces, held together by a domed iron rivet. It has been argued that spectacles lengthened the working life of craftsmen and were thus of considerable economic significance.[7] We do not know of developments in labour saving devices. Labour was cheap and plentiful – except after the period of the Black Death – and the idea of replacing labour by machinery had not been mooted yet.

The main role of technology was, as ever, to support needs, rather than to create or stimulate new demands. The mechanism of stimulating demand by technological innovation was still in its infancy and applied only to a few luxury goods; most stimulation came from trade. The affluent citizen, and certainly the higher clergy and manorial owners and officials, acquired a taste for luxuries, including a better class of furniture and more luxurious accommodation. As in all generations and periods, weapons were high on the agenda of needs. Perverse-

7 David Landes. (1998). The Wealth and Poverty of Nations. London: Little, Brown &Co., pp 46–47

ly, weapons were also used as fashion accessories. Large and small wars raged throughout the Middle Ages, including the crusades, the Hundred Year War between England and France (1338 to 1453) and many wars between rival city-states in Italy. Warfare, as always, led to constant improvements in weapons. Castles and fortresses were built and siege machinery was further developed. In Britain, the Normans established their rule with a series of castles that offered safety to garrisons that controlled the countryside.

Of more lasting importance was, of course, the introduction of firearms into Europe in the 14th century. The first cannon fabricated from forged iron was made in Germany in 1325. In 1350 the cast bronze cannon came into production and by the end of the century this was replaced by the cast iron cannon. Though the older types of artillery, consisting of a variety of catapults, was not immediately replaced by cannon, nevertheless the eventual victory of cannon over catapult was inevitable. The early cannon was difficult to load, difficult to fire, and often exploded in the face of its own crew. The range and accuracy also left a lot to be desired and it took a great deal of theoretical and practical development work to make the cannon into the devastating weapon of the 20th century. A variety of smaller firearms, the harquebus, the musket and the pistol followed in the 15th century.

It is not possible to summarise a long historic period, even from the point of view of the interactions between technology and society. In the earlier part of the Middle Ages wealth was very clearly defined as the ownership of land and power and wealth were, as always, inextricably connected. Those who owned land virtually owned all the people who lived on their land and all the labour they could provide. As the people were extremely poor, their needs could be fulfilled by relatively simple technology. All that was required were the means to till the heavy soil and to clear forests and dry marshes to gain more agricultural land. The heavy plough that was gradually developed from Roman ploughs, and better methods of breeding and harnessing draught animals, were almost all the technology that was required. Labour was cheap, so there was no incentive to develop labour saving devices. Housing and food were minimal and required little or no technological development. Only the building of large manor houses and of churches required more sophisticated building methods and their interiors demanded some good quality carpentry and other competent tradesmen. Most needs were provided locally and the total volume of trade was, initially, very small. War was a constant feature. Most neighbouring kingdoms, principalities, and dukedoms were involved in territorial or other disputes. The armoured knight was a development from Roman cavalry and required highly skilled armourers, good quality steel, good saddlers, and the breeding of heavy horses. The ownership of land became closely associated with military service. This is not surprising. As the ownership of land was the source of power, and power inevitably was involved in power struggles. Theoretically, all power and all land was in the hands of the king or other ruler, but the ruler had to devolve power and handed over land to his followers and associates who, in their turn, provided him with the armies he needed and even with the administration that was required to keep the country running. The feudal lords and their immediate entourage and family were the knights who provided the cavalry and the military leadership of either conscripted serfs or mercenaries.

Technology was entirely pragmatic. No theories were involved. All technologies were developed by trial and error, and all skills and knowledge were passed on from master to apprentice or from father to son. All theoretical thinking was dominated by church dogma or by folklore and prejudice. All formal thought and formal learning was in the hands of the clergy and the monastic orders. Dogmatic theory lends itself to speculation, but does not encourage systematic observation and the construction of theories based on such observation. A mind that is trained in dogma and speculation cannot possibly pursue what we now regard as science. Science is based on careful observation and deliberate experimentation and on the construction of theories designed to explain the observations and results of experiments in the simplest and most logical way. No theory is raised to the status of an article of faith; all scientific theories can, and often are, falsified by further experiment or observation and may be discarded and replaced by a better theory. By better we mean in better agreement with observed facts, simpler, more general, and more capable of predicting the results of future observations. In complete contrast with religious truth, scientific truth is always on probation, valid only until falsified.

In the later Middle Ages, trade developed and trade led to major changes. First, trade needed centres to operate from, and thus towns and cities grew and multiplied. Secondly, trade became an alternative source of wealth. Successful merchants could become very rich and kept their riches in cash, gold, houses, jewellery, and other luxuries. To some extent, they also invested in land and became a new class of non-feudal landowner.

Trade was not, initially, driven by technological developments. It was driven by the realisation of enterprising and ambitious men that money could be made by buying short and selling long. Perhaps the ultimate reason for the development of trade was that land clearance reached its natural limits and thus the need for more agricultural labour diminished. A growing population needed new sources of employment and income. No doubt the efficiency of agriculture increased by gradually improved methods, and the land became capable of feeding more mouths that did not contribute directly to the production of food. This provided an opportunity for more men to leave the land and engage in trade, professions or crafts. As wealth increased, more people required not only food, but also products of technology, such as houses, furniture, clothing, bedding, pottery and other domestic equipment. Traders naturally required transport for themselves and, more importantly, for their goods. Cities required not only dwellings, but also city walls, water supplies, streets, shops, public houses, and a whole range of administrative and other services. All these activities required technology but did not call for much technological innovation. Technology developed slowly and gradually to fulfil the demands made upon it; it did not stimulate many, if any, new demands.

The most visible and lasting heritage of the Middle Ages are the splendid cathedrals built during that period. They are marvellous feats of technology and a testimonial to the ingenuity and skill of medieval stonemasons. Cathedrals were not, however, a demand stimulated by technology, but an expression and technological embodiment of the power of the church and the piety of the people.

The Main Literary Sources Used for this Chapter

Agricola, Georgius (1912). De Re Metallica, translated from the first Latin edition of 1556 by Herbert Clark Hoover and Lou Henry Hoover. London: The Mining Magazine.

Coulton, C. G. (1947). Medieval Panorama. Cambridge: The University Press.

Dyer, Christopher. (2000). Everyday Life in Medieval England. London: Hambledon and London.

Heer, Friedrich. (1961, paperback 1998). The Medieval World- Europe 1100-1350. London: George Weidenfeld & Nicolson, paperback: Phoenix.

Holmes, George (Ed.). (1988). The Oxford Illustrated History of Medieval Europe. Oxford: Oxford University Press.

Oakeshott, R. Ewart. (1960). The Archaeology of Weapons – Arms and Armour from Prehistory to the Age of Chivalry. London: Lutterworth Press.

Pirenne, Henri. (1936, sixth impression 1958). Economic and Social History of Medieval Europe. London: Routledge & Kegan Paul Ltd.

Postan, M. M. (1972). The Medieval Economy and Society, London: Weidenfeld and Nicolson.

Ramelli, Augustini (1620). Schatzkammer Mechanischer Kuenste. Translated into German and published by Henning Gross, Leipzig. See also Ramelli, Agostino. (1976). Various and Ingenious Machines. English translation by Martha Teach Gnudi, Baltimore: Johns Hopkins University Press.

Rawcliffe, Carole. (1995). Medicine and Society in Later Medieval England, Stroud: Sutton Publishing.

Singer, Charles and E. J. Holmyard and A. R. Hall and Trevor I. Williams (Ed.). (1956). A History of Technology, vol. II, The Mediterranean Civilizations and the Middle Ages. Oxford: Clarendon Press.

CHAPTER 5

The Industrial Revolution

The period of rapid change from manual production to machine production, and from cottage to factory, in the 18th and 19th centuries is usually referred to as the industrial revolution. In view of later major technological upheavals, it is now often referred to as the first industrial revolution. The dating of this period, as of all historical periods, is arbitrary and imprecise. Some historians like to extend the period from 1750 to 1900; others prefer somewhat shorter periods. The industrial revolution started in England and was confined to England during its first phase, from about 1760 to about 1830. It was a period when invention chased invention and when large, mechanised, textile and steel industries were built up. Coal and steam provided the energy to drive the new machinery and, hence, coal mining expanded dramatically. It was also the beginning of the railway age, with all its economic and social repercussions, not to mention the military implications. The social changes accompanying the technological change were of far-reaching importance. A new class arose: the working class or proletariat. The middle class expanded by the addition of owners and investors, managers, and professionals based on industry. The new factories drew workers away from the countryside into new or growing cities. Factories provided work of sorts and the landless poor followed their call. Industry provided new opportunities for entrepreneurs and for investors, as well as for higher echelons of workers: managers, engineers, and all the professional and commercial services that go with new or vastly enlarged population centres.

The following summary of the main features of the industrial revolution should provide a general overview of what it was all about. The three main technological developments were:

1. The rise of the factory system of production, particularly in the textile industry;
2. The development of the steam engine as a prime mover and, as a consequence, the development of the railways and an increased demand for coal;
3. The development of iron smelting that made it possible to substitute coke for charcoal and led to a huge increase in iron and steel production. This development is associated with a further increase in demand for coal and the development of town gas.

The associated social developments were:

1. The rise of the industrial city;
2. The rise of an industrial workforce and thus a working class or proletariat;
3. A vast expansion in total consumption of manufactured goods.

The era ushered in by the industrial revolution has not really ended. We still live in an industrial society, though many commentators prefer to call our age post-industrial because industry no longer provides the main source of employment. However, we still produce in factories; we still use power, though now mostly electric power, to drive production machinery, and consumption of manufactured products has continued to expand. Agriculture has continued to lose workers, albeit no longer to industry but to the service sector.

The intellectual climate had changed radically between the end of the Middle Ages and the dawning of the Industrial Age. It all began with the Renaissance and continued with the birth of modern science in the 17th century and the Enlightenment of the 18th century. The 16th, 17th and 18th centuries may be viewed as a continuous period of intellectual change from dogma to freedom. Whereas at the beginning of the period church dogma reigned supreme and intellectual activity outside these strict boundaries was non-existent; at the end of the period intellectual activity ranged freely over all aspects of human activity. Philosophy, natural science, medicine, and art had all thrown off the shackles of dogma and allowed the spirit of enquiry to range freely. The Renaissance, the birth of modern science and of Empiricism, the Enlightenment, all these are phases and movements contributing to freedom of thought, to the deposition of the church from its position as the sole arbiter of all truth and, ultimately, to a leading role for rationality and science. It did not happen without a struggle and the process has never been completed. In fact in these days of religious fundamentalism and fanaticism, one might think that we have been moving backward in recent years.

At the beginning of the period the church still reigned supreme and fought hard to retain its supremacy. One of the toughest battles was that for the central position of the Earth. To the church, this was of supreme importance because the Earth was the centrepiece of God's creation and thus had to be the centre of the universe. The Greek philosopher Aristotle (384 to 322 BC) postulated that the Earth had to be stationary and could not rotate. The later Greek astronomer Ptolemy (2nd century AD) devised an ingenious astronomical model in which the known stars were fixed to a system of crystal spheres revolving in complex motions, with the Earth at the centre of the spheres – the geocentric universe. The church accepted Aristotle's writings and Ptolemy's theory as incontrovertibly correct and not open to doubt. When astronomers of the 16th century, led by Nicolaus Copernicus (born c.1473), suggested that the Earth might not be the centre of the universe but was merely one of several planets revolving round the sun – the heliocentric universe - the church fought long and hard to preserve it geocentric view. The first publication suggesting the heliocentric system was De Revolutionibus by Nicolaus Copernicus, published in 1543. Brave men of science fought protracted battles; sometimes ending in the execution of the scientist branded a heretic, against a church that refused to accept new scientific thinking and interpreted its own dogmatic view of the world in the narrowest possible way. Galileo Galilei (1564–1642) regarded as the founder of experimental science, spent years battling the church and had the narrowest of escapes from extreme punishment, because his writings, though couched in most cautious language, were regarded as supporting the Copernican view. Galileo, one of the earliest scientists to use a telescope for astronomical observation, had concluded from his observations that the Ptolemaic theory was untenable and that the much simpler Copernican theory ought to be accepted. After protracted battles the church eventually accepted that science and religion could be reconciled, though many think that, at best, they can only co-exist peacefully, without real compatibility. This was not Newton's view. Isaac Newton (1642–1727) stressed repeatedly that his motivation for scientific research was the wish to fathom the ways God had constructed the world and thus learn more of God's will. Not many scientists are driven by this motivation in the present day.

Giants of science, beginning with Galileo and perhaps culminating with Isaac Newton, succeeded in showing that science could lead to an understanding of natural phenomena by using experiment and observation as the foundation for theoretical thought and mathematical argument. Science refuted the view, held by Greek philosophers, that logical argument alone could furnish scientific knowledge; knowledge had to be based on observation, and observation had to be the arbiter of correctness of scientific theory. In consequence of this changed intellectual attitude, science developed by leaps and bounds. Newton laid the foundations to calculus, showed that white light consisted of a mixture of light of all possible wavelengths (and thus colours), and discovered both the law of gravity and the fundamental laws of motion. These achievements were made possible by previous experiments and theories of motion (mechanics) by Galileo and the wave theory of light proposed by Christian Huygens (1629–1695).

Though science and technology were not nearly as interwoven and as interdependent as they are in our modern day, I think that there was a great deal of interaction even in the days of nascent experimental science. Technology enabled experimental science to advance and scientific knowledge helped technology to develop. Technology had progressed sufficiently to provide the equipment necessary for scientific experimentation. Lenses and mirrors could be ground for the construction of telescopes and microscopes; prisms could be manufactured to investigate the nature of light, and machinery could be devised to test properties of materials and of gases. Astronomy made great strides in the wake of the introduction of the telescope into astronomical observation, mainly owing to Galileo and to Newton, but also several others. Indeed by mentioning a few names one does great injustice to the many important scientists who remain without a mention. Machines could be constructed with greater confidence owing to the knowledge of the laws of motion on the one hand, and knowledge of the properties of materials owed in those days to Robert Hooke (1635–1703) and his law of elasticity. Biology began to make great strides when Antonie van Leeuwenhoek (1632–1723) built a microscope and discovered the existence of bacteria. Finally, I shall mention Robert Boyle (1627–1691) and Blaise Pascal (1623–1662) who discovered the fundamental laws of the behaviour of gases. This was important, perhaps crucial, for the construction of steam engines.

This amazing bunching of scientific genius in so short a time is rather surprising and calls for some, albeit speculative, explanation. I think that two factors were involved.

First, the church had given up its resistance to scientific thinking. The scientific trailblazers, such as Nicolaus Copernicus and Galileo Galilei, as well as the philosophers and scholars of the Renaissance, had shaken off the intellectual shackles of the church and had achieved a degree of independence of thought and of teaching and learning. Secondly, the general level of economic and social development had brought into being an educated and affluent middle class of merchants and professional men who had sufficient leisure, and sufficient means, to follow intellectual pursuits. It was no longer necessary for almost everybody to engage in agriculture in order to feed the population, and the landowners were no longer the only wealthy and influential class. Universities had been founded in which such men could pursue their researches without intellectual or financial dependence.

Finally, the Enlightenment, though essentially a movement of the 18th century, had truly started in the late 17th century and was, in part at least, a movement for education of a large lay public. Societies for reading and learning and a variety of discussion groups sprang up everywhere and formed a popular counterpart to the founding of learned societies. The learned societies, that were to become prestigious and illustrious institutions, predated the discussion groups and were founded in the second half of the 17th century. The Royal Society of London for the Promotion of Natural Knowledge was founded in 1660 and given a Royal Charter by Charles II in 1662. Of equal importance was the French Academy of Sciences (Académie des Sciences), founded in 1666. Huygens was among the founding members, as were Rene Descartes, Blaise Pascal, and many other illustrious men. One of the moving spirits behind the foundation of the Academy was Jean-Baptiste Colbert, at the time controller general of finance. This shows the involvement of the state as a benefactor of science and proves that since that crucial period science has not only been tolerated, but also actively encouraged. The Prussian Academy of Sciences was not far behind. It was founded in 1700 under King Frederick I and its first president was the famous philosopher and mathematician Gottfried Wilhelm Leibniz (1646–1716).

"For the German philosopher Immanuel Kant, Enlightenment was mankind's final coming of age, the emancipation of the human consciousness from an immature state of ignorance and error"[1]. The encyclopédists Diderot and d'Alembert were of major influence in the Enlightenment, along with many other philosophers in several countries. A full list would be tantamount to a who-is-who of the time and I shall name but a few for general orientation: Montesquieu, Voltaire, Locke, Hume, Rousseau, Kant, Herder, Benjamin Franklin and perhaps Thomas Paine. The Encyclopédie defines the *philosophe* as one who 'trampling on prejudice, tradition, universal consent, authority, in a word, all that enslaves most minds, dares to think for himself.'[2] "...Christianity was ever after with its back against the wall, forever trying to accommodate itself to new knowledge. The Enlightenment by contrast eagerly seized upon the excitement of the infinite."[3]

It was an age of readiness for experimentation and new ventures. There was a new curiosity about the world, which gave rise to new scientific theories and to new mathematics. And there was a new willingness to experiment with new ideas in engineering and in economic enterprise. The scientific knowledge and the intellectual climate were ready for the scientific and the industrial revolutions. And a new bourgeois class was ready to become the social carrier of revolutionary change.

Although the great inventors of the day were not men of science but men of practical experience, this is not to say that they were unaware of scientific knowledge. They may not have been formally educated, but they were

1 See Roy Porter, (2001), The Enlightenment, Palgrave, p. 1
2 Ibid, p. 3
3 Ibid, p. 67

self-taught, knowledgeable and imaginative men. Both during the industrial revolution and in the preceding period that we might describe as the scientific revolution, interchange between scientists and inventors was active. Many private discussion groups made important contributions and made sure that knowledge was disseminated, deepened by discussion and kept alive as a social enterprise.

Unfortunately, not all exchanges among scientists were friendly and constructive. Some extraordinarily acrimonious disputes are known. Robert Hooke and Isaac Newton had many a quarrel, both as to the validity of their scientific theories and as to priority of discovery. The most famous dispute of all was between Newton and Leibniz. The main bone of contention between them was priority in the invention of the calculus. It seems certain that both men had in fact invented it independently in slightly different form. The dispute was as bitter as it was protracted and was never resolved by the disputants.

The spirit of free enquiry and free enterprise spread throughout Europe and America and so did the Industrial Revolution. Though England made strenuous efforts to keep the new inventions and the new skills confined to England, that effort was doomed to failure. Other countries began to imitate the English success, sometimes with the aid of English entrepreneurs. The first country that followed the English example in building up large steel, coal, and textile industries was Belgium. France took a little longer, as in the early phase of the industrial revolution it was in the throes of political revolution and of war. Germany also proved tardy in the beginning and did not start to develop fast till after the unification of Germany in 1871. More countries followed: the United States, Netherlands, Sweden, Austria-Hungary, Italy and, especially after the Bolshevik revolution, the Soviet Union. By the middle of the 20th century, almost the whole world, including Japan, China and India, was industrialized.

Many conditions have to be fulfilled for such major technological change to take place in such a short period, indeed for the change to begin and feed on itself and gather pace.

The first condition is the existence of an efficient agriculture, capable of feeding a growing proportion of the population that is not actively engaged on food production. Only an efficient agriculture can free sufficient labour to enable industrial production to grow. Associated with this condition is the possibility for the rural population to move freely into the growing industrial centres, to provide the new industrial labour force. In feudal times, when agricultural workers were prohibited from moving, industrialisation was not possible, unless the landowners had undertaken it. Considerable improvements in the efficiency and productivity of agriculture indeed preceded the Industrial Revolution.

The second condition was a sufficiently large and developed pre-industrial manufacturing base. Without such a base, there would not have been the reservoir of skills and experience necessary for embarking upon the Industrial Revolution. Fairly large quantities of textiles were produced on simple hand operated machines. A variety of machinery, including forges, pumps, bellows and flourmills, were driven by water- animal- or wind-power. Substantial quantities of iron, copper, brass, bronze and other metals and alloys were produced and a variety of objects were manufactured from them. The pottery industry operated on quite a large scale. Building and civil engineering had progressed to quite a high level and transport was reasonably well developed. The production and consumption of non-agricultural goods had reached a substantial level. In association with all these developments, many skilled craftsmen had been trained in a variety of industrially relevant trades.

The third condition was the existence of a class of people who had capital at their disposal and were flexible enough and mobile enough to try their hand at a variety of enterprises. Traders and merchants who had accumulated some capital and were looking for new commercial opportunities were as important as bankers who were looking for profitable investments and profitable possibilities of lending out their money. It was necessary that the newly rich, and perhaps some of the older rich, were prepared to risk their money on entrepreneurial enterprises and did not aspire to live a life of leisure and luxury. They were prepared to spend money on uncertain enterprise, rather than squandering it all on building palaces, acquiring land, and consuming luxury items such as jewellery and silverware.

Finally, it was necessary for a number of craftsmen to be not only skilled, but also generally educated, intelligent and imaginative enough to see new technological possibilities that might improve existing products or improve the ways existing products were manufactured. In short, a class of inventors and technical entrepreneurs was required.

To recapitulate: the industrial revolution was possible only because agriculture was efficient and could spare labour; the rural population became mobile; there was an adequate accumulation of capital that was available for risky enterprises; there was a sufficient reservoir of skills; and there were some highly gifted and imaginative craftsmen who provided the inventions for the new age.

It is said that the entrepreneurial spirit was particularly strong in England as opposed to, for example, the predilection toward luxury among the French elite of the time. Some argue that the Protestant faith had something to do with a propensity for hard work, rather than leanings toward excessive luxury and the ornate baroque or rococo styles. The so-called Protestant work ethic and the leanings toward austerity and puritanism, might have contributed to the industrial revolution in England and leanings toward luxury might have contributed to its delayed start elsewhere. The arguments concerning the Protestant work ethic, though often used, are not entirely convincing, though they might contain a grain of truth. What is beyond doubt is that a dogmatic religion that demands excessive obedience from its believers does discourage free thought and an entrepreneurial spirit. Similarly, an excessively hierarchical and dictatorial political system puts a brake on inventiveness and on the taking of entrepreneurial risks. In an oppressive regime it is politically too risky to take commercial or technical risks. Considerations of this ilk apply to the Middle Ages and still apply to some countries or sections of the population in the modern world. They may well be contributing causes to the geographic distribution of modern technological, industrial and commercial success.

The textile industry was the largest of all medieval industries, both in terms of labour employed and in terms of turnover and trade. By the time of the industrial revolution, the industry was producing a large amount of cotton cloth in addition to the traditional woollen and linen cloth. In the decade 1750-59 Britain imported 2.81 million lb (about 1.3 thousand tonnes) of raw cotton annually. A hundred years later, in the decade 1850-59, annual imports had risen to 795 million lb (360,000 tonnes). The textile industry was the one that was first affected by the great inventions and innovations of the industrial revolution. Spinning was the bottleneck of the industry. The weavers were crying out for more yarn that the spinners were hard put to provide. From the technical point of view, spinning is a simpler operation than weaving and on both these counts it is not surprising that the famous great inventions in the textile industry started off with spinning machines.

The first spinning machine was the 'spinning jenny', invented by James Hargreaves in 1766. The spinning jenny was still hand-operated, but instead of spinning a single yarn, it could spin several yarns simultaneously and thus massively increased the output of each worker. The quality of the yarn was not very good and it was suitable only for the weft, not for the warp. The next improvement came from the so-called water frame, a spinning machine driven by water power and producing a high quality of cotton thread. It was invented in 1769 by Richard Arkwright. Arkwright was more than an inventor; he was the first industrialist who introduced the factory system of production that replaced the earlier system of 'putting out' work to small cottage industries. Generally a small merchant put out the work to families based on rural cottages. All members of the family could contribute. The merchant supplied the materials, managed the sequence of processes and marketed the finished products. By 1782 Arkwright employed 5,000 workers in his factory. The third famous name among inventors of spinning machines is that of Samuel Crompton, who invented the spinning mule in 1779. The mule was similar to the spinning jenny, except that it was power-driven and not only spun but also drew out and twisted the cotton yarn simultaneously, thus improving its quality. Crompton's mule, reputedly so called because it was a cross between the spinning jenny and the water frame, found almost universal application, as it enabled the mass production of cotton yarn of high quality. Alas Crompton himself obtained only a minute share of the success because he lacked the money to patent his machine and was cheated out of promised royalties. By 1812 there were 360 mills in operation in Britain, using 4,600,000 mule spindles.

The true importance of these inventions is not simply an increase in the productivity of spinners, but the fact that the whole system of 'putting out' was replaced by a factory system. This was necessitated by the fact that the new machines were expensive and large and needed a source of power, and were thus unsuitable for a small cottage industry. The repercussions of the factory system were far-reaching. Large factories were built that employed thousands of workers. This required major capital inputs and new management skills. It also meant that the workers had to move out of the countryside into the vicinity of the factories, and that only the employed worker earned money while the rest of the family could no longer contribute to the family's income.

It further meant that the worker had to submit to the discipline imposed by machines and by managers and had to learn new skills. Some of the former merchants, who had acted as putters-out, saw organisational advantages in the factory system. By centralising the flows of goods into and out of the factory they gained better control over the goods, and by centralising the workers they gained better control over the workforce. The confluence of large machines and administrative advantage commended the factory system to the capitalist to the point of irresistibility. Finally, it meant that a lot of the input into the textile manufacturing industry now came from outside, for example from the designers and manufacturers of the machinery, though some textile manufacturers, such as Arkwright, made their own. In the longer run, conflicts between the emerging industrial working class and the capitalist/industrialist class were inevitable. Whereas previously nascent capitalism was based on successful merchants and bankers, the bourgeois class was now joined by the industrialists. And what were previously independent small tradesmen/manufacturers became threatened in their existence. Some trades continued to flourish – and indeed flourish to the present day – others had the option of quitting and becoming labourers or, possible only for a very few, to join the class of industrial entrepreneurs.

The power-driven mechanical loom inevitably had to follow the spinning machine. As the bottleneck of spinning was removed, the next bottleneck in textile manufacture, the process of weaving, reared its head. The first successful power-loom was invented by Edmund Cartwright in 1785. It was driven by horses at first, but these were replaced by steam power in 1789. Cartwright's factory was burned down in 1791 by weavers fearing for their livelihood, but the loom continued on its path of gradually replacing handlooms. Though in 1834 there were still twice as many handlooms as power-looms (which says nothing about the ratio of their outputs), by 1850 the power-loom had completely replaced the handloom in Britain. The introduction of power looms was not as rapid as might have been expected. This is explained by two factors. First, the investment took a few years to bring substantial financial benefits and thus needed strong financial muscle as well as foresight. Secondly, the weavers on handlooms were prepared to put up with lower and lower incomes rather than give up their skilled trade and were thus able to compete with the industrially manufactured cloth for quite some time, albeit at the cost of starvation.

The innovations described so far would have left the revolution in textile manufacture incomplete. The bottleneck of spinning was removed by the new steam-powered spinning machines and weaving was greatly accelerated by powered mechanical looms. The increased quantities of cloth still had to be finished to give them a pleasant appearance and touch. The most important of the finishing processes was bleaching, thereby turning the unsightly greyish material white and ready to be dyed and printed. Removing the weakest link from a chain reveals the next weakness. These finishing processes formed a new bottleneck that could only be removed by developments in the chemical industry.

The Swedish chemist Carl Wilhelm Scheele discovered chlorine in 1774, and its bleaching properties provided the solution to the bleaching bottleneck for cotton cloth. The original method of bleaching was to spread the cloth on the ground and expose it to sunshine. Unfortunately sunshine is a pretty scarce commodity in the British Isles and, with rapidly increasing production of cotton cloth, the demand on space for bleaching could not be met. More importantly, the old method of bleaching was hardly suitable for factory production. Thus the discovery of the bleaching properties of chlorine and its various compounds were eagerly seized upon.

Perhaps the most quintessential feature of the industrial revolution was the introduction of steam as a prime mover. Whereas previously machines were either operated by human power, animal power or waterpower, a new prime mover became available that was far more powerful than human or animal power and, in contrast to waterpower, was not restricted to locations near suitable streams.

Steam power has many antecedents. The physicist Denis Papin demonstrated the pressure steam can exert and suggested some kind of steam engine with a piston moved by the pressure of steam. This never progressed beyond a model, and Papin is remembered mainly as the father of the pressure cooker and its safety valve. The concept and magnitude of atmospheric pressure were elucidated by Galileo's assistant and successor as professor of mathematics at the Florentine Academy, Evangelista Torricelli, who first measured atmospheric pressure and by Blaise Pascal, who measured the change in atmospheric pressure as a function of the height above sea level.

The most impressive practical demonstration of the forces involved was arranged in 1654 by Otto von Guericke, an outstanding engineer and successful politician and, it would appear, a brilliant showman. Having invented a mechanical vacuum pump, he evacuated the sphere enclosed between two hollow copper hemispheres, fitted together with an airtight seal. In one of the most spectacular public scientific demonstrations in history, he employed eight horses to attempt pulling the so-called Magdeburg hemispheres apart. Although they were held together merely by the pressure of the atmosphere and were only about 14 inches (35cm) in diameter, the horses failed in their attempt, thus demonstrating the tremendous force exerted by the atmosphere. The designers of the first steam engines made good use of this knowledge and let atmospheric pressure do most of the work of the engine.

Thomas Savery produced a steam-driven pump. The pump consisted of a steam boiler that filled a vessel with steam. When the vessel was cooled so that the steam would condense, a vacuum was created that could be used to raise water into the vessel. As atmospheric pressure can raise water to no more than just under 10m, Savery used high-pressure steam to drive the water from the first vessel into a higher one. It is said that the pump was able to raise water to a height of 80 feet (about 25m). Although Savery's patent, granted in 1698, mentioned "Raising of Water and occasioning Motion to all Sorts of Mill Work by the Impellent Force of Fire", his device was merely a water pump without moving parts (except valves). Pumping water from mines was an important task and Savery's pump achieved this with greater efficiency than all the predecessor, usually horse-driven, devices. Because of the comprehensive nature of Savery's patent, the inventor of the first piston-driven steam pump, a kind of steam engine, John Newcomen, had to cooperate with Savery to implement his idea. Between them, they developed the Newcomen engine of 1712 that could have been used as a prime mover, but was used mainly as an alternative steam-driven pump. The Newcomen engine is known as the "atmospheric engine" because the pressure of the atmosphere provided the actual moving force. Steam was used to fill a cylinder equipped with a piston. The piston was connected to another piston, the latter being part of a piston-pump. When the steam piston had reached one end of its stroke, the cylinder was cooled, the steam condensed, and the atmospheric pressure moved the piston, and the pumping piston with it, to the other end of the stroke. The Newcomen engine was a proper steam engine, as the beam connecting the two pistons might have been used to operate other machinery, but it was mostly used to drive the piston of a water pump. One of the amazing uses for which the pump was sometimes employed was to pump water out of a stream onto a water wheel, with the water wheel operating some machinery, such as textile machines or forge hammers. The intention was to avoid the vagaries of changing water levels in the stream, but the system strikes one as mildly absurd. The advantage was that the Newcomen engine was added to existing machinery and a more dependable operation could thus be achieved without a change in the underlying system. Newcomen engines, with some later improvements by John Smeaton, remained in use for about fifty years.

James Watt is generally regarded as the father of the steam engine that came into general use as the prime mover in all industry for a long period. The steam engine could be used to supply dependable mechanical motion in situations where waterpower was not reliably available and thus made it possible to drive industrial machinery wherever industry found it desirable. The electric motor and the steam turbine began to compete with the steam engine in the late 19th century and eventually rendered it obsolete during the 20th century.

James Watt was a mathematical-instrument maker at Glasgow University and this fact made him into a symbol linking the pre-industrial craft traditions with the scientific foundations of modern engineering. Watt the craftsman and the friend and collaborator of university scientists; Watt the inventor, industrialist and Fellow of the Royal Society (from 1785), forged the symbolic link between craft-based and science-based engineering. One of his friends was Joseph Black, who had established the concept of latent heat[4], and whose knowledge was certainly relevant to Watt's endeavour. While repairing a model of a Newcomen engine in 1764, Watt was struck by its inefficiency. While thinking about the causes of the inefficiency, he became aware of the wastefulness of condensing steam in the cylinder itself and thought that it would be more efficient to have a separate condenser, connected to the cylinder by a system of pipes and valves. The invention of the separate condenser was one of the most important steps on the way to the Watt engine. Using a separate condenser

4 Latent heat is the energy associated with a change of state, such as from water to steam.

meant that the massive cylinder did not have to be alternately heated and cooled, thus saving a great deal of energy and reducing fuel consumption of the engine by about 75%. Watt took great care to keep the cylinder hot and even surrounded it with a lagged steam jacket. Though the second law of thermodynamics had not been formulated yet, Watt realised that an engine working between a permanently hot reservoir (the cylinder) and a permanently cold reservoir (the condenser) was the most efficient way of constructing a heat engine.

The full theoretical support for this approach was formulated, particularly in the second law of thermodynamics, by several scientists during the 19th century. The best-known contributors to this fundamental law are Sadi Carnot, Rudolf Clausius, William Thomson (later Lord Kelvin), Max Planck and Ludwig Boltzmann. It can be formulated in a variety of ways, but as far as our interests are concerned, the most important formulation, owing to Carnot, says that the efficiency of a heat engine, operating between a hot and a cold reservoir, increases with an increasing difference in the temperatures of the reservoirs. One of the consequences of the second law of thermodynamics is that the construction of a perpetuum mobile, i.e. an engine that produces work without consuming energy, is not possible. Another formulation of the law, owing to Clausius, states that heat cannot travel of itself from a colder body to a hotter body or, in other words, if we wish to transport thermal energy from a cold to a hot body, e.g. in a refrigerator, we must invest energy.

In 1768 Watt entered into partnership with John Roebuck, the Birmingham industrialist and manufacturer of sulphuric acid, who helped him to build the full-scale improved steam engine. In 1769 Watt took out a patent for "A New Invented Method of Lessening the Consumption of Steam and Fuel in Fire Engines." The next few years saw little progress on the fire engine, as he called the steam engine, because Watt had taken up work as a surveyor in Scotland and Roebuck went bankrupt. In 1774 Watt moved to Birmingham and Matthew Boulton, the owner of a small metalworking factory in Soho near Birmingham, took over the business side of the enterprise from Roebuck. His partnership with Watt was one of the most fruitful ones in the history of collaboration between a businessman and an inventor. The first two Boulton and Watt engines were built and installed in 1776, one for pumping water out of a colliery, the other for working the bellows in a metal smelting furnace. In the following five years several engines were installed in Cornwall for pumping water out of copper and tin mines. One of the many problems encountered in the production of steam engines was the boring of sufficiently accurate cylinders that would prevent the escape of steam. The solution lay in the use of a boring machine invented by John Wilkinson for the production of cannon. This is an early example of cross-fertilization – a method from one branch of technology finding a use in a different branch.

Boulton was keen to expand the market for steam engines and thought that this could best be achieved if the engine could drive rotating machinery, such as in textile mills or corn mills. Watt solved the problem of converting the reciprocating motion of the steam engine into rotary motion by the invention of the so-called sun-and-planet gear, suggested by his assistant William Murdock, patented in 1781. He would have used a crankshaft but for the fact that a patent for this had been taken out by one of his assistants and Watt did not wish to be involved either in litigation or the payment of royalties. The sun-and-planet gear served till 1794, when the crankshaft patent expired. Several further improvements perfected the Watt steam engine. The first was replacing the single working stroke by two working strokes, with the steam being introduced on alternate sides of the cylinder, thus avoiding an idle stroke. The second was using the expansive properties of steam by letting it into the cylinder only at the beginning of the stroke. Watt improved the connection between piston rod and beam by introducing a three bar parallel motion arrangement; an invention he was particularly proud of. Next, Watt introduced a flywheel that smoothed out the motion of the engine and, finally, he used a centrifugal governor of his own design, which kept the speed of the engine constant by controlling the amount of steam let into the cylinder. Interestingly, the five hundred or so Boulton and Watt engines that were produced and sold during their partnership, lasting from 1775 to 1800, all operated with low-pressure steam, only a few pounds above atmospheric pressure. Watt was vehemently opposed to raising the pressure because he apparently regarded high pressure and high temperature as too great a hazard. Had he known the second law of thermodynamics, not formulated yet, he would probably have attempted to raise the temperature and pressure of the steam used, regardless of the problems involved.

After the expiry of Watt's master patent and his retirement in 1800, the Watt engine was built by several engineers in Britain, on the Continent and in America. Many further improvements were made over the years.

The drop-valves, for example, were replaced by simpler and lighter sliding valves. Much of the timber used in the construction was replaced by cast or wrought iron. Richard Trevithick increased the pressure of steam up to 10 atmospheres (145 lb/sqin or 10^6 pascal) and Oliver Evans did the same in the United States. They did this without the benefit of knowledge of the second law of thermodynamics, simply because their intuition told them that high pressure would be more effective than low pressure. Trevithick applied his steam engine to a vehicle, building the first steam locomotive in 1804. The locomotive was able to run on a road but not on a cast iron track because the track cracked under the weight of the engine. The road locomotive was ahead of its time and did not prove a success. Another important improvement to the steam engine came when Arthur Woolf patented the high-pressure compound engine in 1804-05. The principle of compounding, originally suggested by Jonathan Hornblower in 1781 but not put into practice because the Watt patent precluded it, consists of using the steam from one cylinder, after it had expanded, as the input into a second cylinder, working at lower pressure. Further cylinders can be added and the principle became particularly effective when high-pressure steam was used to start with. The condenser followed, if at all, only the last cylinder so that, in contrast to early engines, atmospheric pressure did only a minor part of the work of the engine. Further improvements continued and eventually most steam engines disposed of the huge beam and operated with horizontal cylinders instead of the previous vertical ones. They became sleek, smooth and efficient and operated well into the 20th century.

There are two reasons for telling the story of the steam engine in some detail. The first reason is its inherent fascination. So many separate streams of knowledge and experience that came together in a few fertile minds to produce a machine that changed the world. Knowledge of the properties of steam, knowledge about atmospheric pressure, knowledge about the needs of mines and of other machinery, and knowledge about construction materials and their properties all had to come together. Steam power was an essential pre-requisite for the development of a large-scale mass-producing manufacturing industry and a fast high-capacity system of transport. For better or for worse, the world would have been a very different place without it.

Secondly, the steam engine is a prime example of the process of technological innovation, and the case serves us as an illustration from which we can generalize. Innovation occurs when a technological possibility converges with a market opportunity. The several inventors of the steam engine all saw that steam could be used to power a machine that could perform useful tasks, such as pumping water, driving production machinery, powering winding gear and, ultimately, powering vehicles. Their perception of the technical possibilities was closely matched by their perception of market opportunities. There was no need for market assessment or market research; it was obvious for all to see that steam power, with its greater flexibility and independence of location, would supplement and largely supplant the hitherto common waterpower. It was also plain that the use of mechanical machinery, such as the new spinning and weaving machines, would greatly expand. Apart from that, the steam engine created a market for itself in that it increased demand for coal, and thus the need for more water pumping capacity and more winding gear as well as, in later consequence, for more transport facilities provided by steam-ships and railways.

The first models of an entirely new device are invariably somewhat crude when compared to later models. In other words, a new invention undergoes a lengthy process of gradual improvement. Improvement continues from innovation to obsolescence, but is particularly rapid during the early years in the life of a new technology. When comparing a late 19th century steam engine with an early Newcomen engine, we see that the change owed to gradual development was enormous. Many improvements – or even innovations – are achieved by cross-fertilization, i.e. by using a technology from another field and applying it to the new field. The centrifugal governor, or speed regulator, for example, is said by some historians to have been used previously for the control of millstones when Watt adapted it, or perhaps re-invented it, for use in the control of steam engines. Another example, mentioned earlier, is the use of a cannon-boring machine for the production of cylinders for steam engines.

Generally, new technology depends on previously available technology. When the inventor perceives a new technological possibility, he takes into account the manufacturing facilities of the day. Part of the assessment of the feasibility of a new technology is indeed consideration whether it is possible to produce it with known materials and existing machinery. On the other hand, once an innovation comes into use, the facilities for producing it improve at the same time and as part of the development of the innovation. When the first steam engines were manufactured, woodworking facilities, metal pipes, soldering techniques, screws, rivets, metal working facilities, a variety of valves, steam boilers, the production of rope and chains, bronze bearings and a good many other techniques were part of the current technological repertoire and were all used to produce early steam engines. An innovation has to be perceived within the framework of current technological know-how and the successful inventor, in contrast to the writer of science fiction, has to be aware of the limitations imposed upon his imagination by technological reality.

Generally speaking, the mechanical revolution needed to be complemented by an analogous chemical revolution to complete the industrial revolution. By corollary, the new physical sciences needed to be complemented by new chemical sciences. And indeed the old alchemy soon gave way to a new chemistry.

The story of a development as complex as that of modern chemistry is not one with a proper beginning, middle and end. We begin at an arbitrary point in 1661, with an outstanding contribution by Robert Boyle, who pointed out that the elements postulated by Aristotle and the alchemists need not be indivisible and might indeed consist of mixtures of chemical entities. The fundamental laws of the behaviour of gases and their chemical reactions were subsequently established, mainly with contributions from Robert Boyle, Jacques A. Charles (1746-1823) and Joseph-Louis Gay-Lussac (1778-1850). Of equal importance was the gradual establishment of the idea that matter was conserved and could neither be created nor destroyed in chemical reactions. Two of the contributors to this idea were Joseph Black (1728-1799), whom we have met as a friend of James Watt, and Antoine-Laurent Lavoisier (1743-1794). With experiments on combustion and oxidation by Lavoisier and the discovery of oxygen by Joseph Priestley (1733-1804) and, independently and a little earlier, by Carl Wilhelm Scheele (1742-1786), the process of combustion was finally understood and the fanciful old theory of phlogiston was thrown out of the window. Many elements and compounds were discovered in rapid succession and a logical chemical nomenclature, used to this day, was established.

Vague atomic theories of matter were formulated in antiquity, but they had neither predictive nor explanatory value. John Dalton (1766-1844) was the first to formulate the atomic theory of matter in a way that has proved useful and has survived the test of time. When taken in conjunction with the work of Amedeo Avogadro (1776-1856) and others, we obtain a picture of elements consisting of atoms, with the atoms of different elements having different characteristic atomic weights. A number of individual atoms, of the same or several different elements, can combine to form molecules. Thus compounds consist of molecules made up of atoms of the elements forming the compound. This very rough and very brief and very unfair outline, leaving out much important work, shows that by the beginning of the 19th century chemical theory had advanced sufficiently to act as a guide to practical progress. Whereas physics had been short on both theory and experiment until the days of Galileo Galilei, chemistry had gathered plenty of empirical knowledge in the hands of the alchemists, but had been short on useful theory that would enable further progress.

It is not my intention to describe the complete chemical counterpart to the mechanical industrial revolution. I shall give only a brief and incomplete description of some of the important chemical/industrial developments of the time. One such group of chemicals was known as alkali. The term is a little ill defined, but basically it means the soluble hydroxides of the alkali metals (such as sodium and potassium). They are strong bases and are used in a variety of industrial processes, such as the production of paper, glass, soap and others.

Wood was a well-established raw material for the early chemical industry, if that is an appropriate term for the small-scale production of a few industrially useful chemical products. Wood was used to produce tar and pitch for the shipbuilding industry, and wood ash was also an early source of alkali. The production was expensive and soon became unable to supply the increasing demand. The first substitute for wood was seaweed and kelp. The alkali obtained from the incineration of seaweed indeed played an important role in Britain till about 1830. Kelp was not a pure alkali, but it was workable as such and, in later years, it provided a useful source of iodine.

The most attractive raw material for alkali was common salt, but duties on salt in Britain held back development in this direction. The duty on "foul salt" was revoked in 1781 and a number of processes for the manufacture of alkali were patented in the following years and several firms started manufacture. Eventually, the mainstream alkali industry in Europe, including Britain, was founded upon a process invented by the Frenchman

Nicolas Leblanc in 1787. In this process, common salt was decomposed by sulphuric acid, producing sodium sulphate and hydrogen chloride ($2NaCl+H_2SO_4 \rightarrow HCl+Na_2SO_4$). The sodium sulphate was mixed with coal and limestone and heated in a crucible. The resulting ash was leached with water and the soda ash (sodium carbonate, Na_2CO_3) was obtained by evaporation of the water. The process was patented in France in 1791 and Leblanc built a factory for the manufacture of soda ash, but the events of the French Revolution deprived him of both his patron and his factory. In England manufacture of sodium carbonate by the Leblanc process was taken up by James Muspratt in 1823, first in Liverpool, then in Cheshire. Two years later Charles Tennant took up the process in Glasgow. Initially, the gas hydrogen chloride, one of the products of the first stage of the Leblanc process, was allowed to escape through the chimney and caused enormous damage to the environment. There was an outcry against this devastating pollution and a successful demand for intervention by the law. The process was changed and the chlorine was recovered and used to manufacture bleaches in various forms. There was a growing market for bleaches in the textile and paper industries and thus compliance with the social demand to reduce pollution led to profitable new products.

We have seen that sulphuric acid is one of the starting points for the production of soda ash, but it is also the starting point for many other products and processes. Thus the manufacture of sulphuric acid was an important component of the chemical industry. The substance was known to the alchemists and was first manufactured in Germany by two different processes, of which the oxidation of sulphur by heating it with potassium nitrate (KNO_3) in a glass bell was the more successful. As metal workers in Birmingham used sulphuric acid for various purposes, John Roebuck (Watt's later partner) and Samuel Garbett built a plant for its production in Birmingham in 1746. Roebuck, a highly educated and well informed scientist, knew that sulphuric acid does not attack lead and substituted a lead chamber for the glass bell. The substitution of this robust industrial material for fragile laboratory glassware established the basis for the large-scale manufacture of chemicals. Sulphuric acid came to be used as a basic ingredient in bleaching, printing and dyeing processes in the textile, paper and glass industries. Though the production of alkali was the most important market for sulphuric acid up to the first half of the 19th century, this demand was later outstripped by the needs of the chemical fertilizer and other more modern industries, including the manufacture of explosives.

Since the 1780s it was known that the element phosphorus played an important role in plant growth. Guano and bones were imported and ground to provide both nitrogen and phosphorus fertilizers. In about 1840 Justus Liebig, a leading German academic chemist, found that bone meal treated with sulphuric acid was more soluble and therefore more readily absorbed by plants. In 1842 John Bennet Lawes patented the manufacture of the fertilizer "superphosphate" manufactured by treating phosphate rock with sulphuric acid. This was the beginning of the production of chemical fertilizers that rapidly became a large worldwide industry.

The old dyes for textiles, all derived from plant materials, provided a small range of colours and were not colourfast. Synthetic dyes provided a much greater range and better colourfastness. The father of synthetic dyestuffs was August Wilhelm Hofmann who taught in England for a time, before returning to Germany. One of his students, William Henry Perkin, discovered the first synthetic dye, mauveine, in 1856 and began manufacturing it a year later. The German industry, however, with firms such as Badische Anilin & Sodafabrik (BASF) founded in Mannheim in 1865, and with strong theoretical backing from rapidly developing academic organic chemistry, soon reached a virtual monopoly position in the production of a growing range of synthetic dyestuffs. The main raw material for synthetic dyestuffs was coal tar, a by-product of the coke industry that previously had been more of a nuisance than a welcome product. Another German firm, founded for the production of dyestuffs by Friedrich Bayer in 1863, has since become a world-famous name. Bayer was the first who achieved success with the development of synthetic drugs in introducing Aspirin in 1899. By going beyond the period usually regarded as the industrial revolution, extending the discussion briefly to the end of the 19th century, we hope to obtain some valuable insights as this period marks the beginning of organic chemistry.

Apart from the beginning of pharmaceutical research and the production of new therapeutic drugs, the period saw several major breakthroughs in the advance of medicine. The first was the establishment of the theory that the putrefaction of wounds, some diseases, and the fermentation of milk or beer, were all caused by microorganisms. Though there had been previous speculation on these matters, some dating to antiquity, it was Louis Pasteur who finally established the theory and found the means of preventing fermentation by heating (pasteurizing) the milk and thus killing the relevant microorganisms. The process of pasteurization has found widespread industrial application. Advances in organic chemistry brought with them the discovery of antiseptics. The application of antiseptics by surgeons massively reduced the death rate of surgical operations. One of the pioneers of antiseptics was the English physician-surgeon Joseph Lister. Lister used antiseptics for dressing wounds from 1865. The result was that between 1865 and 1869 the mortality in a surgical ward fell from 45% to 15%. Another major benefit of chemical advances was the discovery that certain substances, such as chloroform and ether, could be used as anesthetics. When Queen Victoria accepted chloroform from her doctor John Snow (the same who discovered the cause of cholera) in 1847, the use of anesthetics spread rapidly. Finally, we must mention vaccination. Pasteur himself introduced vaccination against anthrax and rabies. Edward Jenner discovered a method of vaccination against smallpox in 1796 and his method formed the foundation of widespread immunization programmes. The discovery of x-rays by Conrad Wilhelm Röntgen in 1895 completes our highly selective and brutally abridged list of medical breakthroughs between the end of the 18th and 19th centuries. The discovery of x-rays made the body transparent to a certain degree and thereby opened up tremendous diagnostic possibilities. Later it was discovered that irradiation posed severe dangers and, as the reverse side of the same coin, offered therapeutic possibilities against malignant tumours.

We leave the realms of chemistry and medicine and turn to metal-working and machine tools. The successful inventor combines contemporary technological knowledge with imagination and a vision of the technical future. The inventor and/or his business partner also use their imagination to guess the probable market demand for their invention. To bring together a technical idea with possible markets for it, thus realising an innovation, requires knowledge, imagination, faith and perseverance.

The gradual replacement of timber by iron (or other metals) in the construction of machinery was made possible by improvements in the accurate shaping of iron parts. Product innovation often brings in its wake advances in production technology. In modern parlance, product and process innovation are often closely linked. Generally speaking, the change from small-scale individual hand production to large-scale power-driven mass production, brought about by the industrial revolution, required many advances in machine tools. Simple hand tools had to be replaced by machine tools that achieved greater speeds of production, greater accuracy, and greater consistency and uniformity. The accuracy of machine tools makes parts interchangeable and permits the mass assembly of complex artefacts. The 18th and 19th centuries saw the development of a shaper, invented by James Nasmyth, to produce accurate metal surfaces, a milling machine, invented by Eli Whitney (1765-1825), to produce accurate more complex shapes. Whitney invented the milling machine in his bid for the mass production of muskets with interchangeable parts. In a public demonstration reminiscent of Otto von Guericke, he showed that muskets could be assembled by picking parts at random from a pile. Henry Maudsley developed a metalworking lathe from the earlier simple woodcutting lathe, and developed the first screw-cutting lathe. Up to then, screws had been filed by hand and it was neither possible to produce large quantities nor standardized threads. Each individual screw fitted only its own individual nut. From Maudsley's early machine automated screw-cutting machines were developed and standardization was introduced. An early standard, used to this day, was designed by the outstanding toolmaker of his day, Joseph Whitworth (1803-1887), and is named after him. Both Maudsley and his pupil Whitworth produced accurate measuring devices, a precondition for the accurate production of metal parts. All these inventions benefited the development of the steam engine and, of greater importance, were essential for the development of mass production. It would be difficult to overestimate the role of accuracy in mass production. The mass production of accurate parts to be assembled into a finished product forms the basis of all modern production of complex machinery, from the sewing machine to the automobile, and from the gas-turbine to the robot. It was this approach to production that reduced the price and expanded the market for an untold number of machines and devices in modern use.

We should add a brief speculation on the role of patents. Presumably Boulton would not have invested so much capital into Watt's invention had the latter not been protected by patent, and thus showed promise for profit. Too much early competition would have killed the goose that was about to lay the golden egg; it would have deprived Watt of his just deserts. On the other hand, Watt's monopoly slowed down the introduction of some improvements, such as high-pressure compound engines. But did that harm society? I do not think so.

And yet again, the patent protection of the crankshaft spurned Watt on to develop the sun-and-planet gear and thus gave an impetus to a new invention. Perhaps the verdict on the role of patents must be a rather lame "it all depends" on circumstances and on points of view.

After this detour, we shall return to the main inventions of the industrial revolution and briefly mention the development of the steam locomotive, the steam-ship and railways. The story is so well known that extreme brevity will suffice here. The two best-known contributors to these developments are George Stephenson (1781-1848) and Isambard Kingdom Brunel (1806-1859).

Stephenson acquired his knowledge of steam engines when he got a job as an operator of a Newcomen engine working in a coalmine. He rose to a position of chief mechanic in a colliery and had opportunity to see the operation of an early steam traction engine for hauling coal, built by John Blenkinsop. Blenkinsop equipped the wooden track with a toothed third rail and the engine operated by engaging a ratchet wheel with the toothed third rail. Stephenson regarded this unreliable mode of propulsion as unnecessary and designed, in succession, two successful locomotives with smooth driving wheels. The second used part of the steam to increase the draught of the funnel and thus increased the performance of the engine. Trevithick had used a similar system earlier. After building several more locomotives for the colliery, Stephenson managed to convince the promoters of a proposed horse-drawn railway link between Stockton and Darlington in Northern England to try out a steam locomotive. In 1825 the *Locomotion* pulled the first ever passenger train at 15 miles per hour (24km/h). The line was mainly used to carry coal and proved very successful in bringing down the price of coal. Four years later, Stephenson's *Rocket* won, with a speed of 36mph, a competition to build a steam locomotive for a railway line connecting Manchester and Liverpool. Neither Stephenson nor the steam locomotive ever looked back and a veritable railway fever broke out, building railways and all that goes with them all over Britain, the Continent and America. The unstoppable process had its enemies. The main opposition came from farmers, who feared the loss of their markets for oats if steam engines replaced horses. Opposition also came from citizens frightened of the huffing and puffing smoking monsters that moved at unprecedented speeds.

The development of railways had far-reaching social consequences. The capacity for moving goods increased enormously and this was a pre-condition for expanding industry and trade. The mobility of people increased equally enormously, with consequences for trade, for travel and tourism and for city centres. A vast building programme of railway track, railway stations, bridges, rolling stock, locomotives, and hotels was set in motion and the concept of tourism emerged. City centres were in turmoil as stations and hotels were built and track was laid. New occupations came into being, such as engine drivers, firemen, signalmen, stationmasters. Trade and occupations associated with horses and carts became much reduced, while the new occupations rose. It would theoretically be possible to draw up a balance sheet of profits and losses. No accurate assessment is known to me, but general opinion has it that the gains were greater than the losses. Small consolation for the losers. Some losers were among those who had flocked to invest into the new technology. There was much over-investment and many failures, though many a fortune was also made. The phenomenon of over-investment in promising new technologies is a general one. The bursting of the modern 'dot.com' bubble is not dissimilar.

The military potential of the railways for the quick movement of troops, supplies and equipment was soon seized upon. Particularly the continental nations regarded the railways as of too great a strategic significance to be left to private enterprise and the states became heavily involved in the control and ownership of the new system. In Britain, it was left entirely to private enterprise, which resulted in some chaotic development.

Isambard Kingdom Brunel became chief engineer of one of the new railway companies, the Great Western Railway, in 1833. The construction of railway lines involves a great deal of civil engineering, including the building of bridges and tunnels. Brunel made considerable contributions in this field, as well as in the design of railway stations. He was an engineer of remarkable versatility and apart from his contributions to the advancement of railways and of civil engineering, also designed the first ocean-going steamships. Before Brunel, small river steamships had been in operation ever since Robert Fulton, with considerable commercial and political backing, had built the *Clermont* that operated on the Hudson river from 1807 and was followed by a great number of paddle steamers on the great American rivers. The history of this invention is controversial but the controversy is of no great interest to us. The first of Brunel's three ships, the *Great Western*, launched in 1837, had a wooden hull and was propelled by paddle wheels, but also carried sails. It was the first steamship to provide a regular transatlantic service. His next ship, the *Great Britain* (1843), had an iron hull and was the first large ship with a screw propeller invented by John Ericsson. The case of the ship's propeller is interesting from the point of view of historiography. Whereas the Anglo-Saxon histories of technology ascribe the invention in the 1830ies to the Swede Ericsson (and occasionally to the Englishman Francis Smith), German and Austrian historians ascribe the invention to the Austrian forester Josef Ressel and date it to 1827. The final of Brunel's steamships, the *Great Eastern* (1858), carried the principle of "belt and braces" to an extreme. It had a double iron hull, two paddle wheel engines, two propeller engines, as well as six masts rigged with sail. It was originally built for the Eastern Navigation Company to operate on the route to India and was the largest ship in the world; too large for most ports. The ship was sold to different companies and eventually was employed to lay the first successful telegraph cable across the Atlantic, completed in 1874.

Steamships were soon developed, built and operated by a great many companies and came into universal use for the carriage of goods and passengers. The Great Western Steamship Company, owners of the first two of Brunel's ships, was not commercially successful and was soon overtaken by a shipping company founded by Sir Samuel Cunard that became known as the Cunard Line. Competition between British, American, French, German and other shipping lines became fierce. The compound steam engine propelled them at ever increasing speeds. The overall impact of steamships upon world trade cannot be overestimated. World trade grew rapidly and much of this growth was owing to the increased speed and capacity of shipping. After the collapse of the revolutions in Europe in 1848, the desire for emigration to America increased and the steamship made a considerable impact in providing a large capacity for carrying emigrants with great discomfort at the cheapest possible rates. As usual, the old technology fought back and made improvements in an attempt to avoid or delay its demise in the face of the new technology. Sailing vessels improved greatly both in speed and in capacity, culminating in the fast clippers, but were unable to withstand the competition from steam and had virtually disappeared by the end of the 19th century.

A qualitative and quantitative leap in the production of iron and steel is another hallmark of the industrial revolution. Steel became, and has remained, the most important material of human civilization. Machines, machine tools, buildings, means of transport, all this and much else became dependent upon iron and steel. Although we like to think of our present age as the silicon age, in reality steel is still the most important material that sustains our civilization. The first step in raising iron production was made by substituting a plentiful fuel – coal converted to coke – for the increasingly scarce charcoal used hitherto.

Forests are among the first victims of any rising civilization and industrialization.[5] Timber is cut down for building, shipbuilding and construction, as well as for charcoal and for heating and cooking. Forests are cleared for pasture and agriculture. It is not surprising that charcoal was in short supply when increasing demands were made and many forests had disappeared, especially in Britain. One of the properties of technology is to seek solutions to problems such as shortages of certain materials or bottlenecks in certain manufacturing processes or in supplies. As charcoal was becoming scarce and expensive and timber supplies were limited by their very nature, many people tried to substitute plentiful coal for scarce charcoal. Technologists and businessmen are always in search of opportunities and any shortage provides an opportunity. Compared to Britain, it took many more years before charcoal was dispensed with in Germany, because the shortage of charcoal was not as acute.

The revolution in iron production started a little earlier than the mainstream industrial revolution. All iron prior to the 18th century was smelted in furnaces fuelled with charcoal. As charcoal became increasingly scarce and expensive in England, many unsuccessful attempts were made to use plentiful coal instead. Success came in 1709, when Abraham Darby, a cast-iron ware manufacturer in Coalbrookdale (Shropshire, England) succeeded in producing pig iron in a blast furnace fuelled with coke. The advantages of coke are many, quite apart from its cheapness. It is less brittle than charcoal and thus larger furnaces with heavier loads of iron ore and fuel can be built. It also reaches higher temperatures and the iron produced is more fluid and thus Darby's cast-iron products became finer and his business flourished.

---

5 In O. Rackham, The History of the Countryside, Dent, 1986, it is argued that the destruction of forests is to blame entirely on agriculture and not on industry.

Some of Darby's success was owing to the lucky circumstance that the coal he used had a low sulphur content. In general, coke-smelted cast-iron was not as strong as the charcoal-smelted variety and was not as successfully converted to wrought iron by the traditional method of forging. The problem was caused by the presence of impurities in the coke, particularly sulphur and impurities in the iron ore.

The problem of converting poor quality pig iron into good quality wrought iron was solved by a series of inventions, starting with the so-called puddling process, patented by Henry Cort in 1784. The process essentially consists of stirring molten iron with iron rods and thus bringing the impurities to the surface and continuously removing the surface layer. The process is labour intensive and can only be carried out in batches, but it was successful and was used by a large number of manufacturers all over the world. However, the puddling process was too slow to keep pace with increasing iron production from ever-larger blast furnaces and new methods of producing wrought iron were sought. At this stage we shall introduce the term steel in lieu of wrought iron, as it has become customary to use this term for all iron products with a carbon content below 2%. Iron with higher carbon content is still known as cast iron.

The first converter for the conversion of pig iron into steel by a non-mechanical process was announced by Henry Bessemer in 1856. Several others had similar ideas; particularly William Kelly in the USA had the same idea but could not put it into practice for lack of industrial support. The converter is a large vessel that can be tilted and the essential process consists of blowing air through molten pig iron. The air converts the impurities, especially silicon and manganese into oxides that are removed with the slag. The resulting steel was poured out (cast) and was of better quality than the blooms produced in the puddling process. By sheer luck, the iron used by Bessemer was free of phosphorus and it turned out that this was a condition for the success of the original Bessemer process. The problem of phosphorus was solved in 1875 by Sidney Gilchrist Thomas, who used his considerable knowledge of chemistry to suggest lining the Bessemer converter with a strong basic substance, such as dolomite or burned limestone. This would combine with the phosphorus and could then be removed with the slag. He was granted a patent in 1877 and the combination of a Bessemer converter with the Thomas improvement was highly successful in the production of cast steel.

An alternative conversion method was developed by the brothers Wilhelm (Sir William) and Friedrich Siemens and by Pierre-Émile Martin. The essential feature of the open hearth furnace, developed by Siemens, was to increase the temperature of the blast of air by using a heat exchanger, heating the incoming air by the outgoing flue gases. A patent was granted in 1861 and the open-hearth method became more widespread than the Bessemer process. Martin succeeded in producing better quality steel by using the open-hearth furnace, and adding scrap steel to the pig iron. He too obtained a patent, but was financially ruined by patent litigation while firms using his process made fortunes. The process became known as the Siemens-Martin process and came into very wide use for the production of steel. It was essentially replaced only in the middle of the 20th century by the basic oxygen and other processes.

Between 1838 and 1845 Robert Bunsen, a German university professor famous for his Bunsen burner, examined the efficiency of British and German blast furnaces. He made several suggestions that successfully improved the quality of the pig iron and the thermal efficiency of the furnaces. This is an early example of the collaboration between practical engineering and academic science. Good quality cast iron was of great importance as engine cylinders as well as cannon were manufactured from it. In 1779 Abraham Darby built the first bridge made of cast iron in Coalbrookdale and this was the beginning of the era of iron and steel bridges.

The successful methods introduced in the British iron and steel industry were soon transferred to other industrial countries. John Wilkinson introduced the use of coke for smelting iron in a blast furnace to Le Creusot in France in 1782. He also started the production of armaments there, but the industry began to flourish only when Eugène Schneider and his brother Adolphe bought the works in 1836 and expanded them to become one of the world's top players in steel, shipbuilding, armaments, locomotives and so forth.

The first firm to introduce the Bessemer converter (1862) and the Siemens-Martin open-hearth process (1869) on the European continent was Krupp in Essen. The firm was founded by Friedrich Krupp in 1811 and greatly expanded under the leadership of his son, Alfred Krupp. Krupp produced a large range of steel products, among them the first seamless steel railway tyre. The firm became most famous (or notorious) for their production of cast steel cannon, first shown at the Great Exhibition in London in 1851. By the time of Alfred

Krupp's death in 1887 the firm had supplied artillery pieces to 46 nations and he had acquired the nickname Kanonenkönig (cannon king). On the other hand, Krupp was a paternalistic industrialist who created many welfare facilities for his workers, such as a pension fund, housing, hospitals and schools. By the time of his death the firm had grown from employing 7 workers to a workforce of 21,000. The firm played a major role in World War I with its 420mm howitzers, nicknamed Big Bertha, which, jointly with the Austrian 305mm howitzer built by the Skoda works, destroyed the Belgian fortifications in 1914 and enabled the German army to enter northern France via Belgium. Big Bertha could propel shrapnel weighing nearly a tonne over a distance of about 15 km.

The Skoda works are another example of the new engineering industry expanding throughout Europe and America. Emil Škoda was an engineer who bought a small machine workshop in Plzeň in Bohemia in 1869. The town was then located in the Austro-Hungarian Empire and is known as Pilsen in German. Although the Skoda works achieved great importance, the town's world fame happily rests more on its beer than on it guns. Skoda became a major manufacturer of industrial machinery, locomotives, cars, and arms, including armour plate, heavy artillery and, in 1890, one of the world's first machine guns.

We cannot leave the expanding steel industry without at least a brief mention of steel in the USA. The most outstanding name is Andrew Carnegie (1835-1919). The son of a Scottish hand-weaver impoverished and made rebellious by the mechanical looms and forced to emigrate to America, Carnegie started work as a young boy in menial tasks but rapidly rose through the ranks by virtue of his outstanding talents. Having achieved managerial positions and a great income through shrewd investments, he began to concentrate on steel in about 1872. He founded the J Edgar Thomson Steel Works near Pittsburgh, which later became the Carnegie Steel Company. His company was the first to introduce the Bessemer converter in the USA in 1870 and, some years later, the open-hearth process. Carnegie was a leading light in the drive toward greater efficiency and greater vertical integration of the firm. American steel production exceeded that of Britain from about 1890 and the Carnegie Steel Company was the leading producer. Andrew Carnegie retired in 1901 and devoted his time and money to philantropic activities. The various Carnegie foundations are still among the largest and most active philantropic organizations.

Virtually all land and sea transportation became dependent upon steel. The expanding rail network consumed large quantities of steel and probably could not have expanded unless steel replaced cast iron for rails. Steel replaced timber in the construction of textile machinery and in the construction of the very epitome of the Industrial Revolution, the steam engine. Steel converted these machines from slow moving huffing and puffing monsters into sleek smooth machines. Steel products became specialised according to end use, though surprisingly stainless steel did not become commercially successful till the early 20th century. The ready availability and consistent quality of steel contributed greatly to its extended use in the building and construction industries. Steel and steel reinforced concrete became the major construction materials for bridges, dams, large buildings, including skyscrapers, tunnels, and pipelines. Advances in agricultural machines owed much to improved quality and availability of steel and cast iron. Modern industry is built on a foundation of steel. Silicon has been added as another fundamental material in mid 20th century.

It is not possible to single out some achievements as more fundamental than others, but if I had to choose the four most important technological innovations of the Industrial Revolution, I would select: 1. improvements in the manufacture of iron and steel; 2. the introduction of mechanically operated production machinery; 3. the introduction of the steam engine; 4. the increase in accuracy of machining operations. The most crucial organisational innovation, conditional upon the technological innovations, was, of course, the introduction of the factory system of production with all its far-reaching grave social consequences.

All the major and minor mechanical inventions that added up to the Industrial Revolution were made by skilled craftsmen, not by academically trained engineers or scientists. Men such as James Watt had good knowledge of science, but they were not scientists in the usual sense of the word and had no university training. All the inventions of machine tools were made by skilled mechanics, many of them trained by a single outstanding master toolmaker, Henry Maudslay. The inventions in textile machinery and many innovations in steel production were also owing to master craftsmen or self-taught industrialists. The innovations in chemistry and in medicine, as well as later innovations in heat engines, were introduced by academically trained scientists or medics. It would seem that craft training, coupled with natural intuition and talent, were perfectly capable of making mechanical inventions, whereas intuition fails when it comes to chemistry and the more abstract thinking required for more recent technological innovations.

Unfortunately, advancing technology also led to a revolution in arms production and warfare. Gunpowder, the so-called black powder, is said to have been invented in the 10th century in China and used predominantly for fireworks, though its first military use in a primitive bamboo rocket is said to date to the 13th century. The real origin of black powder is disputed, though why so many nations and people should lay claim to this deadly invention is an enigma. Black powder consists of about 75% saltpetre (KNO_3), 15% charcoal and 10% sulphur. For many civilian applications, such as mining or tunnelling, the potassium nitrate is replaced by sodium nitrate. From about 1862, when Alfred Nobel built a plant for the manufacture of nitroglycerin, black powder was gradually replaced by this new explosive and its various derivatives, including trinitrotoluene (TNT). They came into military use from about 1904.

Firearms in the modern sense seem to have made their first appearance in the 14th century. The musket started life as a muzzle-loading shoulder held firearm evolved from the smaller harquebus in the 16th century in Spain. In the middle of the 19th century it was replaced by a breech-loading rifle. Breech loading is much faster than muzzle loading and the term rifle implies a weapon with a rifled bore, i.e. one with a shallow spiral groove cut inside the barrel. Using an elongated projectile, the groove imparts spin to it, which keeps it on a steady course and thus increases the accuracy of the weapon. The rifle became the dominant infantry weapon and, in advanced and automatic form, is in use to this day.

Breech loading became effective when combined with a cartridge made of brass or similar, containing the projectile, the propellant and a percussion cap. The rifle with bolt action, i.e. a spring-loaded bolt that strikes the percussion cap when the trigger is released, came into universal military use by the end of the 19th century. The battle of Königgrätz (now Hradec Králové in the Czech Republic) between Austria and Prussia in 1866 showed the overwhelming advantage of breech loading. The Prussians had an early version of the breech loading bolt-action rifle, whereas the Austrians were equipped with muzzle loading rifles. The Prussian soldiers could shoot six times in the time taken by the Austrians for a single shot. The Austrians and their Saxon allies lost the battle and the war. Thus a relatively small technological advance changed the course of European history.

The most obvious development owing its advance to better materials, especially to cast steel, was the artillery. Cannons became larger and more robust and their range and accuracy increased greatly. Artillery shells became more sophisticated, heavier and more deadly. From about 1850, rifled gun barrels became standard and this necessitated the use of elongated, rather than round ammunition and much greater accuracy was thus achieved. Shrapnel, invented by the British artillery officer Henry Shrapnel, was introduced in the 1790s. Originally it consisted of a spherical shell packed with black powder and musket balls and equipped with a fuse. It had the effect of deadly musket fire of great intensity over a long range. Shrapnel and fuses for different purposes were developed, such as for use against infantry, or against armour.

With the advent of the railways and various other heavy vehicles able to move artillery pieces, armies became more mobile. With the advent of breech-loading rifles and cannon, their firing power increased greatly and warfare changed from man-to-man combat into a longer-range exchange of deadly fire, with only occasional hand-to-hand combat carried out with bayonets on rifles. Only the cavalry still used lances and sabres, but only till the beginning of the First World War. During that terrible war the face of warfare changed again, as it did in every major war fought since the Industrial Revolution.

We may regard the Industrial Revolution as the period of transition between technology serving human needs and technology creating needs with the intent of making money from the gratification of the needs thus created. Up to the Industrial Revolution technology mainly served real vital needs for food, clothing and shelter and, for the majority of people, not much else. It is true, of course, that ever since the earliest kingdoms technology provided luxuries for the rich, but the majority of the population remained untouched by such riches. It is also true that technology provided the means of warfare since the dawn of humankind and, alas, to this very day. I do not wish to argue that warfare is a real need – quite the contrary it is the greatest curse of humanity and the destroyer of all hope for a world worth living in. However, it appears that war is deeply ingrained in human society. Society without war has remained a dream that lives only in the minds of people without power, in rhetoric, and in fiction. We are forced to the unhappy conclusion that technology, in serving the needs of warfare, is serving a vital need of a sick society. Though perhaps we are letting technology off too lightly. Perhaps part of the blame for warfare falls upon technology and technologists and their unbounded love for the creation of ever-new fearful means of destruction. Technologists are not, of course, the controlling factor. Control is firmly in the hands of the rulers who demand the weaponry and finance its development, purchase and use. Technologists go wherever the jobs are and do whatever the higher authorities, be it management or government, demand of them. Technologists are, nevertheless, guilty of feeding the demand with constantly improved weapons and military equipment. They may only be doing their jobs, but do they not show more enthusiasm than is strictly necessary? Scientists and engineers get carried away and become fascinated by their researches and deadly inventions. The world is a complicated place and even those engineers who might be inherently reluctant to be involved in weapons development are often persuaded by arguments of necessity and by the conviction that they merely help to defend their own people. Nobody acknowledges being the aggressor – everybody is the victim of aggression and persuades engineers and scientists that it is their duty to help defend their country against the attacker. The industrial-military complex, this unholy alliance between arms manufacturers and the military, is alive and well and is still a major force for evil in world politics. They have a very large influence on what technologies shall and shall not be developed.

During the Industrial Revolution technology acquired two new properties that mark its transition from servant of need to servant of greed. The new entrepreneurs were in it for the money. They unashamedly invested in technology for profit and equally unashamedly sought inventions that would multiply their profits. Of course the inventors were in it out of curiosity and regarded the tasks they set themselves as challenges to be overcome. But even the inventors kept at least one eye on profits and their motivation was a mix between engineering and financial challenge. The investors, though fascinated by the new technologies, and thus also motivated with a mix of greed and curiosity, were motivated mainly by the hope of profit.

The second new property of technology was the arousal of new needs. When it became possible to manufacture cotton cloth more cheaply, people bought more cotton cloth in the firm belief that they needed it. When it became possible to travel faster, people, at least those who could afford it, discovered that they were in a great hurry to get to faraway places. When the pendulum clock was invented, people of means found it absolutely necessary to furnish their homes with one or more suitably ornate pendulum clocks. And so forth, the list could be expanded and, with further industrialisation and further inventions, it could be expanded to infinity. In our day it is no longer possible to draw up a finite list of artefacts at our disposal. Yet technology is inventing more and more and economists and politicians call, in rare unison, for accelerated innovation. Not many will argue that these innovations are driven by need, rather than by greed. And yet in the developing world (if indeed it is developing) people are malnourished or starving and are forced to drink unsanitary water that causes disease and death.

We call the Industrial Revolution a transition period, because technology only started on its path of becoming a tool of greed and, arguably, still only provided for needs that were not far removed from being essential. Profligate consumption was about to begin, but had not begun yet. On the other hand, the destruction of the environment began in earnest with the Industrial Revolution. As the consumption of coal for heating and for the production of steel and the generation of steam to power machinery rose, so the output of sulphur dioxide increased and began to destroy forests and buildings. Cities began to be covered in soot, smog, grime and smoke and became places of destruction of buildings, plants and human health. Increasing traffic densities caused congestion, accidents and an appalling stench. The nascent industry regarded the rivers as freely available sewers and it has taken 200 years to clean them to a pre-industrial state.

Living conditions in the densely crowded housing of the new industrial labourers were appalling. Working hours were long, working conditions terrible, disease was rife and life expectancy very low. The introduction of mass production did lower the prices of goods and people could, and did, increase their consumption of manufactured goods, but the benefits of the new productive capacity were unevenly distributed and the effects upon urban amenities, and upon the landscape surrounding urban centres, were devastating. In fact many of the newly created industrial centres show scars of indiscriminate development during the nineteenth century to this day. Derelict land, slag heaps and slum dwellings still disfigure many industrial areas.

As the cities grew and cheap dwellings were erected for the newly arrived industrial workers, problems of sanitation became severe. In London most houses were equipped with cesspits only and these were often allowed to overflow. The overflow was washed into inadequate sewers and ended up in the tidal flow of the river Thames. The sewage was washed up and down the Thames and entered into the drinking water. The result was that the whole of London became extremely smelly – it became known as the Great Stink – and that deadly outbreaks of cholera were frequent. It was believed for a long time that cholera was carried in the stinking air, until Dr John Snow discovered in 1854 that a particular outbreak was associated with drinking water drawn from a particular parish pump. By the middle of the 19th century the houses of the well to do were equipped with some form of privy and these were mostly connected to cesspits and occasionally to sewers. Nevertheless, the custom of collecting night soil in cities continued till the end of the 19th century. The poorer districts had communal privies serving several houses.

Parliament debated the issue of sewage for a long time and eventually allocated a large sum to the Metropolitan Board of Works to build a proper sewerage system. The system was built between 1859 and 1875, under the direction of the chief engineer Sir Joseph Bazalgette, and much of it is operational to this day. The sewers were built of brick and the discharge was carried to the East of London. As the flow was driven by gravity, it reached its destination well below the river and had to be pumped to the level of the river with the aid of steam engines. It was discharged into the Thames at high tide and was swept into the sea by the outgoing tide. The episode of London sewerage demonstrates that private investment, such as in cheap housing, cannot cater for the public good unless forced to do so by suitable regulation. Some form of public intervention is essential if the public interest is to be safeguarded.

Two other developments proved of importance in making life in the new crowded cities more tolerable. In addition to providing sewers, the provision of clean piped water was a major step toward healthier and more comfortable living. The arrangements for the supply of water varied greatly from city to city. In 1848-49 nine different water companies supplied London, supplying a total of about 200 million litres daily. When water was simply drawn from the rivers and the rivers became increasingly polluted with industrial outflows as well as with untreated sewage, the risks of disease increased greatly and aroused public alarm. In 1840 the British Parliament set up a select committee to look into the matter, resulting many years later in the establishment of a Waterworks Clauses Act (1847) and a Public Health Act (1848). By this time the water companies were no longer drawing water from the most polluted parts of rivers. Their supplies consisted of a mixture of water drawn from cleaner parts of rivers, from deep wells, and of water brought in from greater distances. From the beginning of the 19th century water filters gradually came to be used. They consisted of layers of washed gravel and of washed sand. The water passed slowly through the filter that had to be cleaned from time to time. In order to cope with changes in demand and in water levels of the rivers, the water companies or municipalities, as the case might be, started a major programme of building reservoirs. The water from the reservoirs was deemed to be clean. By the middle of the 19th century most urban houses had water piped to them, albeit intermittently, forcing them to have their own cisterns.

It is likely that the combined effect of proper sewerage and clean water contributed as much to public health as all the medical sciences put together. Be this as it may, the 19th century saw a great increase in population. The population of England, Wales and Scotland was 10,500,956 in the first census of 1801. By the census of 1851, the population had almost doubled to 20,816,351. Similar increases occurred in other countries. The main growth occurred in the cities, not so much because of increasing population as because of the continued migration from country to town.

After sewerage and water supplies, lighting was the next most important improvement for city dwellers. Before the advent of electric light, gas lighting provided a very acceptable intermediate solution. Though there were many prior investigations and inventions, the story of gas lighting begins in 1801 with a public demonstration by Philippe Lebon in Paris. Lebon obtained gas by destructive distillation of wood[6] in what he called "thermolamps" and burning the gas to obtain both heat and light. In Britain, the most important investigations into the properties of coal gas and its uses were carried out by a leading engineer of the Boulton & Watt Company,

[6] Destructive distillation of wood means heating it in the absence of a supply of air.

William Murdock. In 1792 he produced gas by destructive distillation of coal and used it to light a room. The firm showed little interest in his continuing experiments, until one of Watt's sons alerted them to the possibilities of gas lighting after he had attended the demonstration by Lebon. From 1804 Boulton&Watt sought orders for gas lighting equipment and manufactured it till 1814. Though several gas lighting systems were installed in cotton mills and elsewhere, the business did not fit well into the programme of the successful firm and was abandoned in the face of more successful competition, particularly by Samuel Clegg. Clegg, like Murdock before him, used a cast iron retort to distill the coal but, unlike Murdock, purified the gas by passing it through a lime solution and bubbling it through water, thus eliminating some of the nasty smell of earlier gas. The burners consisted of small nozzles or a narrow slit in a metallic container. Clegg's equipment was installed in cotton mills and other factories and in some large houses.

Despite these modest successes, it was obvious to the entrepreneur Frederic Albert Winsor (a German called Friedrich Albrecht Winzer who came to England in 1803) that producing gas on the spot for each lighting installation was not the way forward. Instead, he suggested producing gas in central generating plant and piping it to consumers. Such enterprise required too much capital to be financed by a single businessman and Winsor tried, with eventual success, to establish a limited company. He found a sufficient number of investors and, in 1812, obtained a charter from Parliament. Previous to the charter he had demonstrated the potential of gas lighting for illuminating streets by installing gaslights in a section of Pall Mall in London in 1807. The company was formed and eventually settled on a new name, The Gas Light and Coke Company. The production of coke and the production of coal gas are much the same process: the destructive distillation of coal. Coal gas consists mostly of hydrogen, methane and carbon monoxide and its production leaves coke and coal tar as by-products. Coke and coal gas were usually produced by the same company in the same plant. After a brief period of erratic management by Winsor, Samuel Clegg joined the company and by the end of 1815, some 26 miles of gas main had been laid in London. The pipes were cast iron, whereas in the USA wooden pipes were used. The connections to the consumers were made with smaller steel pipes that had to be fabricated and welded from strips until some time after the invention of a pipe-drawing process by Cornelius Whitehouse in 1825.

The early lighting was suitable only for large spaces because the combustion of the gas fouled the atmosphere. Matters improved when an atmospheric burner was introduced in about 1840. This had at least two, sometimes three pipes: one for the gas, one for a supply of air, and one for removing the combustion products. Gas lighting became truly competitive with electric lighting when the simple flame was replaced by an incandescent gas-mantle, obtained by surrounding the flame with a fabric soaked in a mixture of rare-earth elements. The device was invented by the Austrian professor of chemistry, Carl Auer von Welsbach, in 1885. In this form gas lighting survived well into the 20th century.

It has been argued that street lighting reduced vice and crime in the cities. On the other hand, it has been argued that the provision of lighting in factories has caused the workforce, including children, to work longer hours. It is very easy to argue that the introduction of new technology caused these social ills. We can indulge in flowery rhetoric claiming that machinery killed innumerable newborn infants in the industrial slums or that murderous machinery enslaved, killed and maimed both child and adult workers. We can draw images of merciless machinery crushing their haggard young victims. In a sense, these arguments are true. But in another sense, it is false to blame machinery for the evils imposed by humans upon their fellow humans.

To manufacture anything at all we need to employ so-called factors of production. We need raw materials, suitable machinery, premises, energy, and so forth. All these are usually subsumed under the term capital. In addition to capital, we need workers – labour - to convert all these inputs into the finished product. The main factors of production are labour and capital and, to some extent, managers are free to choose what mix between these factors they wish to use. There are severe limitations on the choice, but some choice is available. The natural inclination of capitalist owners and their managers is to reduce the labour input as far as possible. Labour is awkward to manage and, more importantly, management wishes to reduce the cost of labour as far as possible. The tendency is to pay the lowest possible wage to the lowest possible number of workers. In the early days of industrialisation the possibilities of reducing the labour input were rather limited because the capabilities of machinery were very limited. The only way to save on labour costs was to pay low wages. The balance between capital and labour has shifted ever since, with machinery becoming more automated and technology developing more and more in the direction of labour saving.

In the days of the Industrial Revolution labour was being forced off the land and thus became freely available for employment in industry. Capital was available because many merchants and other wealthy people had accumulated large amounts of money seeking profitable outlets. Finally, total consumption of manufactured goods was limited by the purchasing power of the population rather than by its willingness to purchase. Markets were not saturated and thus any reduction in price could cause an increase in consumption. The introduction of novel production machinery achieved both increased production and increased productivity. Increased productivity led to a reduction in the price of those items manufactured by the new methods. If everybody can buy as much bread, or meat, or apples or any other food as they are able to eat, then we speak of a saturated market and a reduction in price will make no difference to the amount consumed. Clearly, at the time of the industrial revolution the consumption of items such as cotton cloth was limited by purchasing power rather than by need, and demand rose with decreasing prices.

Thus, at the beginning of the industrial revolution, three essential factors were favourable to increased production: availability of labour, availability of capital, and market demand. Two further factors were needed: availability of useful inventions and of entrepreneurial spirit. The coming together of all these diverse factors created a constellation favourable to the Industrial Revolution.

It needs to be added that the availability of capital is not sufficient; the available capital must be in the hands of people willing to take the risks of starting new enterprises. Similarly, the availability of inventions is not enough if the other conditions for implementing the inventions and turning them into innovations are not fulfilled. An invention is merely an idea; it becomes a technological innovation only if and when it is put to practical use.

The social ills caused by the industrial revolution were owing to the social attitudes of the entrepreneurs. Increasing production and decreasing the prices of items of daily consumption, such as clothing and bedding, are not evil in themselves. Employing surplus labour from agriculture in industrial production is not a social evil either, though whether the growth of cities was desirable is a moot point. What was evil was the fact that working conditions were inhuman because the factory owners were too greedy to pay proper wages and provide better conditions and shorter working hours. What was equally evil was that nobody in power cared about living conditions for the workers or environmental conditions in general. Perhaps it would be too much to expect individual entrepreneurs to be humanitarians under conditions of severe competition. Competition drove wages and prices down and nobody cared that it also drove out all human consideration from the process of industrialisation. But if being humane was too much to ask of the competing individual factory owner, it should have been the business of society as a whole to uphold human values in the face of the brutal onslaught of inhuman technology. When we say society should have taken care of these problems, we inevitably enter the realm of politics. For society as a whole can act only through its governance and the governance of the day was firmly in the hands of the upper classes and these were not interested in anybody's welfare but their own. Looking after the environment to safeguard the planet was not thought of at that time.

The Industrial Revolution was a watershed in the relations between technology and society. The introduction of the factory system of production caused a fundamental change in social relations and caused the production of a large range of products to increase greatly and their prices to fall. As if that were not enough, the Industrial Revolution also changed the status of technology from a provider of necessities to a means of making money. The Industrial Revolution marks the divide between a technology that satisfies needs and a technology that gratifies greed. From the point of view of the consumer, an ever-increasing range of goods on offer aroused needs that did not previously exist. From the point of view of the manufacturer and the technologist, technology provided opportunities for profitable investment and technological innovation expanded these opportunities. The Industrial Revolution started, or reinforced, a trend toward using technology not so much to produce goods that we need, but also as a means of making money out of goods that we can sell despite the fact that they are not strictly needed. Technology thus provides both needs and wants and, occasionally, luxuries as well.

The introduction of certain technologies has some consequences that, in one form or another, are unavoidable and can only be avoided by not using the particular technology. Thus the coal mining industry had to be built up to increase output when the combination of coal-fired boilers and steam power became the prime mover in industrial production. It could have been avoided only if people had been prepared to locate all their productive capacities near suitable sources of waterpower or if people had decided not to use the new production machinery. The former would have meant that only a limited number of production sites would be available and industry could not grow at an appreciable rate. The latter would have meant that industry would hardly have grown at all and that people would have had to forego much of their future consumption of cotton and industrialists, or potential industrialists, would have had to forego most of their future profits. Either of these ways of avoiding industrial growth might seem a good idea to some, but basically humans are greedy and want both more profits and more goods.

If we accept that the total rejection of the new technological possibilities was not a realistic option, we remain with the possibility of using the technology and accepting the unavoidable consequences, but avoiding its most damaging impacts. This certainly was a feasible option and it is a sad reflection upon humankind that it was not used or even seriously considered. Coal mines had to be expanded, but was it necessary to employ child labour or make adult labour work such long hours under the most arduous and dangerous conditions? Migration from villages to industrial centres was unavoidable, but could not the housing provided have been of better quality? The use of land for industrial purposes was necessary, but could not dereliction have been avoided and chimneystacks built sufficiently high to avoid the worst of air pollution? The environment was used as a free for all dump, thus saving direct costs but imposing these costs on the environment and thus on the community at large. From the point of view of the industrialists, the costs were conveniently externalised.

The new machinery, like all major technical change, made a lot of skills redundant. When the new power-driven mechanical looms and spinning machines were introduced, the existing spinners and weavers were, naturally, up in arms. They feared that they would be deprived of their livelihoods by being unable to compete successfully with the more efficient new methods of production. They tried two different ways of resistance. Some became militant and destroyed some new machinery and new factories. They are known as Luddites after their probably mythical leader Ned Ludd. They attacked machines and factories at night, but never attacked people. Only when a factory owner caused the shooting of a band of Luddites, was he murdered in consequence. The movement started in late 1811 and lasted only till 1813, when vicious suppression by the government of Lord Liverpool, including several hangings and transportations, put an end to it. The other path of resistance was no more successful. In a vain attempt to hold their own against the new manufacturers, the established craftsmen accepted lower and lower wages for their work until their meagre earnings became so low that it was impossible for them to continue. The economic case for power looms was not, however, immediately overwhelming and it was not till about 1850 that the handlooms disappeared completely.

To understand the impact of the factory system a little better, we shall briefly follow the sequence of development of textile manufacture. First came the cottage industry, where peasant households produced rough woollen cloth from raw wool to finished cloth. This was largely replaced by the putting-out system, where a merchant bought the raw materials and took them through various stages of manufacture to specialist carders, spinners, weavers, dyers or fullers. The specialist workers were more skilled than their peasant predecessors, but still worked from home and owned their simple machinery. They did not, however, own the materials and did neither buy nor sell but were paid for their labour by the merchant, whether by piece rates or by wages. By the time of the industrial revolution, cotton had replaced wool as the dominant material. The putting-out merchant was a predecessor of the industrial entrepreneur, though his outlay of capital was not nearly as great and neither was the degree of control he exerted over the workforce or the process. The industrial system may, however, be regarded as an extension of the putting-out system. In a sense, it took the putting-out system to its logical conclusion. Instead of taking the raw materials to the various workers in sequence, it concentrated the workers and the sequence of operations in a single place. Instead of using individual small machinery, it concentrated the mechanical process into a sequence of power-driven large mechanical devices. Instead of exerting indirect control over the workers, it took complete direct control over them.

The factory owner is forced to employ a large amount of capital on which he will not make an immediate profit, but which will usually prove profitable in the longer run. He has to deploy hope and faith in addition to money. Whereas the small domestic producer was employed by the putting-out merchant, the latter was not actually on the premises and the worker and his family were in control of their own work schedule and had to deploy their skill and knowledge in the production process; the factory worker was entirely under the control of the managers. The working hours were fixed, the work output and the pace were enforced, the worker had no control over anything and very little skill was demanded. Most of the skill had been incorporated in the design of the machines. Though putting-out required a degree of organisation and exerted a great deal of pressure on the worker, it still left the worker with a degree of discretion and the putting-out merchant with a degree of flexibility. In the factory system, on the other hand, the management had to plan and control every step of the manufacturing process in detail. The flow of materials, from the raw materials and energy supplies right down to the sale of the finished products had to be organised; the machinery had to form a system that had to be maintained in good working order. Workers had to be hired and fired, their wages had to be paid and their work discipline had to be enforced. The worker became a mere cog in a large wheel, his discretion was nil and his previous skills were redundant. The new capitalist became the master of men in ways in which the merchant had not been. The community of industrial workers became the working class, whereas their artisan predecessors had belonged to a class of minor independent producers. The families of textile workers still often combined their industrial work with agricultural activities, as had nearly always been the case in the early cottage industries. The industrial worker under the factory system could not combine his arduous work with any other activity.

Technology creates many problems, but also creates many solutions to the problems it has created. We should modify this statement; it is not technology alone that creates the problems or the solutions, it is the way humans apply technology that is often problematic. Consider the example of the creation of densely populated major industrial cities that were the consequence of the rise of industry. The housing erected for industrial workers was of the lowest possible standard and the highest possible density. Initially, virtually no amenities were provided. No sewerage, no lighting, no running water, no proper toilets, let alone bathrooms. Though the application of new production methods demanded an increase in the working population, it did not directly demand that the new workers had to be accommodated in slum dwellings. In any case, the fact that mass production technologies were introduced was owing to human inclinations of greed, not owing to a technological imperative. However, we must start our chain of argument somewhere, so let us start it with the fact that mass production technology was introduced and that workers had to move from the countryside into industrial areas. Surely housing for them could have been of better quality, less dense, more spacious, with little gardens and with proper sanitation. Street lighting was not invented yet but was introduced rather slowly when it had been invented. Other facilities could have been provided and were not. The sole reason for the creation of appalling slums was the fact that industrial workers were paid low wages and could not afford the cost of higher quality housing, quite apart from probably being overcharged by greedy builders and landlords. The problem of slum housing was thus not created by mass production, but by the fact of low wages. Why wages were low is a moot point. Some argue that there was insufficient productive capacity to pay higher wages that would have meant greater consumption of scarce resources by the workers. Others argue that the low wages were a result of the poor distribution of resources, with the rich consuming more than their fair share and not leaving enough for the poor. We cannot resolve this argument; all we can do is point out that the effects of technology are determined as much by social, economic and political decisions as by the properties of technology.

The density of population in the industrial cities and their lack of proper amenities caused epidemics of cholera and other diseases. The provision of proper sanitation solved these problems relatively quickly. But it is arguable that tuberculosis was also caused by industrialization and the solution of this problem took a very long time. In fact it is not completely solved to this very day. The problems created, at least indirectly, by the application of mass production technologies were alleviated by the application of further technologies. Eventually sewerage was introduced, clean water was piped into houses, gas lighting was provided, toilets were improved, and the streets were paved. Technology, as applied by human society, creates problems; technology often can, if society so wishes, solve or alleviate the problems it created.

The problem of obsolescence of skills was in part alleviated by the demand for new skills. Weavers, spinners, dyers, and fullers became obsolete as their crafts were taken over by machines and methods of mass production. On the other hand, some skills were demanded in the design, production and maintenance of the new machinery and in the development of new processes. New skilled occupations, such as boilermakers, engine drivers, toolmakers and mechanics, were created. We do not know the numerical balance between skills lost and skills created, but we do know that for those craftsmen who lost the opportunity to earn their living by exercising their skill the change was catastrophic. We do know that the number of unskilled workers in the factories, who worked long hours under dreadful conditions, far exceeded the number of skilled workers. We do not know whether the new industrial workers were better or worse off when they exchanged their existence as agricultural labourers for an existence of factory workers. We do know that the change was irreversible. The factories were there to stay and the work in agriculture was gone forever. Technological change traps societies into new situations that can be changed only with the introduction of the next generation of technology or by strong political will and action. The third law of technological change[7] states that such change is mostly irreversible. Though this is not a law of nature, it is nevertheless inescapable.

The newly grown cities, and particularly London, became desperately congested, especially during the period when railroads and sewers were built simultaneously. Horse drawn traffic congested the streets and infested the air. Horse drawn traffic also caused quite a few deaths; the traffic accident is as old as wheeled traffic. The motorcar was hailed as the solution to the dirt and smell and congestion. It has not brought about any solution at all, perhaps the contrary. Technology, unwisely used, cannot solve the problems of previous technology, perhaps equally unwisely used.

Before the industrial revolution, society was largely rural and cities were small. Production was entirely in the hands of artisans working in small workshops. Total consumption of energy and material goods was small. At the end of the industrial revolution, society was industrial. Cities had grown enormously, populations had grown, and large numbers of people worked in large factories. Total consumption of energy and of material goods had increased hugely.

Main Literary Sources Used for this Chapter

Braun, Ernest. (1984). Wayward Technology. London: Frances Pinter Publishers.
Brentjes, Burchard and Siegfried Richter and Rolf Sonnemann (Ed.). (1987). Geschichte der Technik. Köln: Aulis Verlag Deubner &Co KG.
Checkland, S. G. (1964). The Rise of Industrial Society in England 1815-1885. London: Longman.
Engels, Friedrich, ed. David McLellan. (1993). The Condition of the Working Class in England. Oxford: Oxford University Press.
Gay, Peter. (1977). The Enlightenment: an Interpretation. The Science of Freedom. London: Norton Paperback.
Gross, N. T. (1972). The Industrial Revolution in the Habsburg Monarchy 1750 – 1914. The Fontana Economic History of Europe, vol. 4, ch. 5. General Editor: Carlo M. Cipolla.
Hobsbawm, E. J. (1973). The Age of Revolution – Europe 1789 – 1848. London: Sphere Books.
Jones, Howard. (1973). Steam Engines; an International History. London: Ernest Benn.
Landes, David S. (1969). The Unbound Prometheus – Technological Change and Industrial Development in Western Europe from 1750 to the Present. Cambridge: Cambridge University Press.
Lane, P. (1978). The Industrial Revolution. London: Weidenfeld & Nicolson.
Mantoux, P. (1961). The Industrial Revolution in the Eighteenth Century. London: Jonathan Cape.
Mathias, Peter. (1969). The First Industrial Nation. An Economic History of Britain 1700-1914. London: Methuen & Co.
Morazé, Charles (Ed.). (1976). History of Mankind-The Nineteenth Century 1775-1905. London: George Allen & Unwin.
Musson, A. E. (1972). Science, Technology and Economic Growth in the Eighteenth Century. London: Methuen & Co.
Porter, Roy. (second edition 2001). The Enlightenment. Basingstoke: Palgrave.
Singer, Charles and E. J. Holmyard and A. R. Hall and Trevor I. Williams (Ed.). (1958). A History of Technology, vol. IV, The Industrial Revolution, c1750 to c 1850. Oxford: Clarendon Press.
Tann, Jennifer (Ed.). (1981). The Selected Papers of Boulton and Watt. London: Diploma Press.
Toynbee, A. (1906). Lectures on the Industrial Revolution of the 18th Century in England. London: Longman.
Tunzelmann, von G. N. (1978). Steam Power and British Industrialisation to 1860. Oxford: Oxford University Press.

[7] For a summary of the fundamental laws of technological change see ch.7.

Chapter 6

The Twentieth Century

The 19th and 20th centuries form a continuity of technological and scientific development. Technology continued to develop along the path it had taken during the Industrial Revolution. Machine tools of all kinds had become available, the factory system had developed, prime movers and improved materials had been developed, and all these served as a springboard for further industrial and technological developments. The intellectual framework for the 20th century had also been laid by the end of the 19th century. All the sciences had achieved a theoretical framework that serves them to this day. Science had become firmly based on experiment and observation. Theory served to explain and relate observations and predict new phenomena to be observed and results to be obtained from further experimentation. Thus theory formed a framework for understanding and a guide to further experimentation. Any theory that was contradicted by observation would be modified or discarded. Theory could not stand on its own feet as a speculative story; it had to be firmly based on facts. In this way natural sciences became clearly separated from philosophy, theology and humanities. As this separation became established, natural science moved closer to technology until engineering and science became so interwoven that their separation is no longer entirely clear.

Physics and chemistry had reached a high degree of development and served as the scientific foundation of chemical and mechanical industries. Biology had left the era of mere description and, with Darwin and others, had joined the other natural sciences in its methods. Biology and chemistry became the main ingredients of a scientifically based medicine. With anaesthetics, antiseptics, vaccination, scientifically developed pharmaceuticals and properly established studies of anatomy, physiology, pathology and all that, medicine had joined the modern age alongside engineering. University education expanded greatly. Science became a respected occupation rather than just a pastime for gentlemen of means. Technological development no longer was the domain of talented craftsmen but joined the ranks of studies based on science. All these foundations had been laid by the end of the 19th century, and thus the way was clear for the expansion of science-based technology.

Technology had come of age and its importance was recognised by governments and the academic and industrial establishments. Before the 19th century was out, several specialist technological universities (in all but name) had been founded. Examples are the Zurich Polytechnikum[1] (1855); the Massachusetts Institute of Technology (1859); the Technische Hochschule in Berlin-Charlottenburg (1884); and the Faculty of Applied Science in the University of Liege (1893). The University of Cambridge, England, created a chair of Mechanism and Applied Mechanics in 1875 and introduced an undergraduate course in mechanical engineering (tripos) before the end of the 19th century. The French École Polytechnique goes back to the end of the 18th century and over the years has become something of an elite university. Many polytechnic schools, teaching technology somewhat below university level, though often developing into fully fledged technical universities, were founded in the 19th century in many towns and were successful in training a new generation of engineers.

Another indication of the importance assigned by governments to technology is the establishment, in the last quarter of the 19th century, of government laboratories for standards. Their task is to maintain standards in measurement, which is fundamental to modern science and technology. They have expanded beyond their original tasks and now carry out diverse research in science and technology. Examples are the US Bureau of

[1] Later to become the Eidgenössische Technische Hochschule (ETH), highly respected to this day.

Standards, the UK National Physical Laboratory, and the German Pysikalisch-Technische Bundesanstalt (formerly Reichsanstalt).

The twentieth century saw momentous technological changes. To describe them all in any detail would be impossible in a single volume. Fortunately, we do not need a full description for the purpose of this book. I shall restrict myself to very brief descriptions of those technological developments that have affected society most profoundly and shall seek common trends in disparate technologies and their effects. I shall constrain the story, in the main, to the developed countries, for it is in these countries that the effects of technology are felt most directly and most immediately. The developing countries are affected less directly, though forces related to technological developments do affect the problems they struggle with. Regrettably, however, the problems of development, important and difficult as they are, fall without the scope of this narrative.

The choice of technologies to be discussed is, inevitably, an idiosyncratic one. It is very easy to point out a large number of extremely important technological developments excluded from my list and my only defence against such criticism is that it is impossible to be all-inclusive. My list is only a personal choice of examples of important technological developments to be discussed, entirely without prejudice against other technologies that might well be of equal or even greater importance. The examples I wish to discuss are, of course, chosen because they are most familiar to me. To my mind life in the twentieth century was most profoundly affected by the following technological developments:

1. Transport, including motorcars and lorries, fuel supplies and road building, fast trains, container ships, oil tankers and aeroplanes.
2. Electricity, as a pre-condition for information technology and all that, as well as because of power stations and grids, lighting, electric domestic appliances, and electric motors as prime movers in industry.
3. Information technology, including computing, communications, mass media and automation.
4. The chemical industry, particularly plastics.

Some of my omissions may surprise some readers. Notably space exploration and advances in medicine, pharmaceuticals, and food production. I regret these unavoidable omissions. As far as space exploration is concerned, I discuss the importance of near-space, i.e. satellite communications and all that it implies, but consider distant space exploration to have little economic or technological significance. Rockets do, of course, have very great military significance, but scientific exploration of distant space is merely an add-on feature of military research and is of interest to pure science only. I regard the fanciful stories about future migration of humans into outer space as pure entertainment, with no social significance. The social impact of near-space technology, on the other hand, is very great indeed, both in military and in civilian terms. Communications, weather forecasts, navigation, surveying, military intelligence and much more have been significantly advanced by the use of satellites.

We are obsessed with fame; hence we want to know the inventors of technological devices and view them as stars of technology. For the media there can be only one inventor of all important things. Reality, however, is not like that. Complex devices or systems, such as the motorcar, or television, or the computer and its components, are not invented by a single person. There inevitably is a long series of predecessors who lay the foundations for the invention, whether in terms of scientific knowledge or of predecessor technologies. The invention itself also usually has more than one progenitor. Similar ideas strike many people when the time is ripe, i.e. when the technological and scientific foundations have been laid and when social conditions are right. It is naïve to expect a complex machine or system to be invented by a single person. Not only does the same or a similar idea occur to several people at about the same time, but also each inventor needs collaborators to work out the practical details of the invention. In fact recent and current technological innovation has been happening in large teams with a leader who is a primus inter pares rather than a star. Furthermore, before the invention becomes an innovation, i.e. before it reaches the market place, a number of people become involved who are neither scientists nor engineers. They may be financial sponsors, they may be manufacturers or they may participate in the early marketing efforts. When all this is taken into account, who is to say that this or that person brought about this or that innovation? Even in the 19th century the apparent sole inventor was often not so solely responsible for the invention as is popularly assumed. Surely it is a gross oversimplification and distortion of the truth to say that James Watt invented the steam engine. He was a genius and made enormous contributions to the development of the steam engine. But even Watt had many predecessors, collaborators, sponsors, friends and advisers. And even Watt was involved in numerous patent disputes with rivals who, rightly or wrongly, claimed rights to various aspects of the development of steam engines.

Not only do we passionately seek stars to admire and worship, we like to have stars in our own country. The assignment of the title of inventor for this or that technology is often tinged with national pride. Take a relatively simple device such as the sewing machine. Ask any Frenchman and you will be told that the sewing machine was invented by Barthélemy Thimonnier in 1841 in France. Ask an American, and you will be told either that the sewing machine was invented by the American Elias Howe or by the American Isaac Merrit Singer. An Englishman might tell you that the inventor of the sewing machine was James Starley and an Austrian will claim with total conviction that the sewing machine was invented by the Austrian tailor Joseph Madersperger. No doubt there is a Russian inventor too and possibly many others.

The Automobile

The most visible and most obvious impact of 20th century technology upon modern society is that of the automobile. There are many claimants to the fame of having built the first petrol driven automobile. One of the best-supported claims is that of Carl Benz, who built a practical four-wheeled motorcar in 1893. A few years earlier, in 1885, Benz had built a three-wheeled motor vehicle with a single cylinder two-stroke petrol engine. By 1888 he was employing 50 people to manufacture the three-wheeler. Gottlieb Daimler and Wilhelm Maybach built the first high-speed internal combustion engine and mounted it on a two-wheeled vehicle in 1885, probably the first motorcycle. One year later they built a four-wheeled motorcar. However, even though most of the major inventions had been made in the closing years of the 19th century, it was the 20th century that saw the motorcar becoming a practical proposition and finding widespread application.

Before the motorcar driven by a petrol engine, several steam-driven vehicles made their appearance and, as soon as the combination of electric motor and electric battery had become practicable, some electrically driven automobiles came on the market. Among the builders of steam-driven road vehicles Amédée Ballée seems to have been one of the early successes with a series of vehicles built in the 1870s, though of course Trevithick was earlier and the principle of steam driven locomotion had by then been well established in railway locomotives. Although Telford completed his great London to Holyhead road in 1830, powered road transport was slow to take off in Britain. Whereas railways were spreading rapidly and a veritable railway fever had spread, early powered road transport spread faster in France than in England. Whereas the British were wary of the new vehicles and gave them a hard time by the legal requirement (from 1865 to 1896) that a man with a red flag should walk in front of such a vehicle to warn pedestrians and horses of the oncoming monster, the French upper classes were much more inclined toward accepting powered road vehicles than their British or German counterparts. In real terms this man with the red flag, so beloved in folklore of motoring history, had very little effect upon the development of the motorcar in Britain. At the time when development began in earnest, the law had been rescinded. Far from looking upon the motor vehicle as an object of everyday use, in its earliest days it was considered a source of fun. A plaything for the rich, which allowed them to display both wealth and manliness. Motor vehicle trials and races started in France as early as 1896 and became a regular feature that contributed much to the evolution of the motorcar.

Many early motor vehicles owed much to the technology of the modern bicycle, invented in 1885. The bicycle became extremely popular and its frame made of steel tubes, its chain-drive, and its ball bearings all proved useful in car manufacture. Thus early motor vehicles were closely related to horse-drawn carriages with bicycle technology and, of course, an engine added. Many of the early motor manufacturers were originally bicycle manufacturers, among them Opel, Peugeot, and Rover.

The logic of invention of the motorcar is easy to construct. Steam engines were the first method of converting heat into mechanical energy and motion. The locomotive was the first result of putting a steam engine on wheels and using the engine to propel the locomotive and let it pull further wagons loaded with goods or people. The rails offered the great advantage of reducing friction between the wheels and the substrate and thus reducing the power requirement. The idea of putting the steam engine on something akin to a horse-drawn carriage was a fairly obvious one and the first steam-driven automobiles started from there. They became more practical when steel tyres, used on carriage wheels, were replaced by rubber tyres, albeit solid ones at first. The next logical step was to do away with a separate boiler. Surely a more compact and more efficient engine could be designed if the heat was not produced in a separate boiler and then fed to a cylinder in the form of steam. If the combustion process could take place in the cylinder itself, driving the piston more or less directly, higher efficiency should be achieved and much space and weight could be saved. Thus the internal combustion engine was thought of and transformed into reality. The mounting of such an engine onto a bicycle, tricycle or four-wheeled vehicle was then almost a matter of course.

All the required predecessor technologies were available. Cast iron and steel of good quality and the machinery to shape these were available. Steel tubing, rubber tyres, chains, valves, tappets, and all the other technologies were in place. The only thing that was missing was market demand. The first motorcars were the clear result of a technological dream and the hope that sufficient numbers of people would be willing to share this dream. The car was a fulfilment of a technological possibility and the hope was that rich men would find the idea of riding in a motorcar, at what was seen as high speed, quite irresistible. The car was not meant for the masses, it was not meant to solve a practical problem, in fact it was not seen as an item of utility but an item of fun; a toy for the rich. And so it was for quite a long time, especially in Europe. In a very real sense this trend is being revived today, in the early years of the 21st century, when several new ultra-luxury, ultra-expensive, ultra-gas-guzzling vehicles are brought onto a market that is saturated with universally owned mass-produced cars of all shapes and sizes. The ultra-rich are getting back their toys and their means of demonstrating their wealth. The current financial crisis and environmental concerns might, hopefully, soon put an end to this madness.

It is not surprising that one of the first uses (if that is the right word) of motorcars was racing. Only the very rich and the manufacturers of motorcars could afford to participate. For the rich it was fun, for the manufacturers it was a sales gimmick and a development programme. The first automobile test was run from Paris to Rouen in 1894 and the first road race from Paris to Bordeaux in 1895. 16 petrol-driven four-wheel vehicles, 7 steam vehicles, 2 electric carriages and 2 motorcycles participated in the race. Nine cars finished within the prescribed 100 hours, one steam car and 8 petrol-driven ones. The winner was Emile Levassor, driving a Panhard-Levassor powered by a Daimler 4 horsepower engine at an average speed of 24 km/h (15mph). The brothers Michelin used the first pneumatic tyres in this race. Their car finished the race, but not within the prescribed 100 hours and not without a great deal of trouble with tyres and everything else.

The most frequently acknowledged inventor of the modern motorcar gasoline engine is Nikolaus Otto. Otto built his first internal combustion engine in 1861. In 1864 he formed a partnership with the industrialist Eugen Langen and together they built the first four-stroke gasoline engine in 1876. This engine was awarded the gold medal at the great Paris exhibition of 1876. The partnership lasted till Otto's death in 1891 and may be compared to the many years of fruitful cooperation between the industrialist Boulton and the inventor Watt. The Otto engine used the four-stroke principle first described by Alphonse Beau de Rochas and patented by him in 1862. Though Beau de Rochas first described the cycle, it has become known as the Otto cycle because it was Otto who first built a practical engine based on the four-stroke principle. Otto and Lange were commercially rather successful with their slow-running, often gas-burning, engine that replaced many a steam engine in mainly stationary uses. The firm sold 50,000 engines in the first 17 years of operation.

The theoretical foundations to de Rochas' work had been laid previously by Sadi Carnot. Carnot, an army officer and trained engineer, was disturbed by the fact that French steam engines were extremely inefficient compared to their British counterparts. He started thinking about the problem of engine efficiency in an abstract way and formulated a general theory of heat engines. He showed that the efficiency of an ideal heat engine, irrespective of the operating cycle or fluid used in it (steam, gas or an air-liquid fuel mixture), depends only on the temperature difference between the hottest and the coldest part of the engine. He designed the so-called Carnot cycle, an ideal operating cycle for an engine operating between a hot and a cold reservoir[2]. All real cycles, whether for steam engines or internal combustion engines, try to come as close as possible to the ideal

[2] The Carnot cycle consists of a reversible isothermal (constant temperature) expansion and compression and of a reversible adiabatic (constant entropy) expansion and compression.

Carnot cycle. His work came to the attention of the engineering profession through the efforts of Émile Clapeyron from about 1834.

Gottlieb Daimler and Wilhelm Maybach built the first ultimately successful high-speed petrol engine. An essential aspect of this engine was a newly designed carburettor, the device that allows liquid fuel to be mixed with air to form a gas-like compressible combustible mixture, invented by Wilhelm Maybach. Maybach's patented carburettor became widely used from 1893 and, as so many lucrative inventions, became subject to patent litigation. Daimler and Maybach formed the Daimler-Motoren-Gesellschaft for the production of automobiles in 1890. Karl Benz formed his own successful company in 1883 that built automobiles from 1885. The Benz Company merged in 1926 with the Daimler Company to form the hugely successful Daimler-Benz Company. More recently, the company has merged with Chrysler of the USA to form one of the few surviving major global motor manufacturing companies. The marriage between the two partners ended fairly quickly in divorce and Chrysler is looking for a new bride.

Another contender for the crown of inventor of the modern motorcar engine, and indeed the modern motorcar, alas a less successful one, was the Belgian Etienne Lenoir. Lenoir may lay claim to having built the first motor car with an internal combustion engine in 1862, though his vehicle was wholly unpractical and took two or three hours to cover a distance of six miles. There are many more inventors who claim to have built the first motorcar, and their claims are often supported more for nationalistic than for historic reasons. In America, for example, there are three contenders for the title of inventor of the gasoline-powered motorcar. The matter is one of indifference to me for the purposes of this book; I have no wish to act as umpire in matters of priority of inventions.

The next most important development in the design of internal combustion engines also falls into the closing years of the 19th century. Rudolf Diesel patented the design of the engine named after him in 1892 and built the first practical single cylinder four-stroke diesel engine with an output of 25hp in 1897. His design was the result of a determined attempt to design an engine that would run on a cycle as close as possible to the ideal Carnot cycle. Because of its high efficiency and, hence, frugal consumption, the engine became an immediate success for stationary, marine, locomotive and heavy vehicle applications. It took many more decades of development before it became an accepted alternative to the Otto engine, first for smaller goods vehicles and, eventually, for passenger cars. The modern Diesel engine is indeed a highly efficient engine that has captured a large segment of the automobile market. It has also become competitive with the gasoline engine in the environmental "friendliness" of its exhaust gases,albeit only in connection with particle filters that remove the carcinogenic particles of soot from the exhaust gases.

The emerging motor industry demonstrates what appears to be a fundamental law of new industries: the law of initial expansion followed by concentration. When the first motorcars were built and knowledge about them began to spread, every entrepreneur with access to basic mechanical workshop facilities started building motorcars. A bandwagon got rolling and a large number of people jumped on it. The number of car manufacturers rose rapidly during the final years of the 19th and the early years of the 20th century. Cars were hand-made individually; no standard design had emerged and the technology was readily accessible to a large number of people with some mechanical expertise. Patent protection proved entirely inadequate. Though there was much litigation about patents and even Otto lost his patent rights when a prior claim by Lenoir was upheld, it was not possible to stop people building individually designed cars and engines. By 1898 about fifty companies in America were building automobiles and this number rose to 240 in the next few years. The number soon began to decline and by about 1930 there were only three major manufacturers, five important independent manufacturers and a small number of also rans. Most of the new upstarts sank without a trace, but among the survivors are many of today's household names of motorcar manufacturers.

The initial expectations for car sales were, by modern standards, very modest indeed. The car was seen as an extremely attractive toy for the rich rather than as a universal means of transport, though the possibilities of building motorbuses and military transporters were envisaged quite early. Despite a limited vision of market openings, it was confidently expected that sales would be sufficient to keep quite a few manufacturers happy. The initial barriers to entry were not too difficult to overcome. Patent protection was weak and the required manufacturing facilities were widely available in all kinds of mechanical workshops. Those were ideal conditions for the setting up of many companies and the bandwagon got rolling. Engine manufacture was more complex and many a small maker of cars bought in engines from a much smaller number of engine manufacturers. However, with further technological developments, competition became tough and only the biggest, luckiest and most efficient players survived, while the small fish went under.

What went wrong? What does always go wrong when a technological bandwagon gets rolling? It happened with railway companies, it happened with domestic appliances, it happened with television sets and it happened with electronics and with computers. Why? There can be several causes. One is saturation. As the initial potential of the market for first time buyers is exhausted, only the replacement market remains and this may be insufficient to keep the many firms who jumped on the bandwagon in business. When every potential customer has, say, a motorcar, only the replacement market remains. This can be stimulated and kept artificially high by so-called built-in obsolescence, but it is still somewhat limited.

The saturation mechanism operates in many industries, but in the case of the motorcar industry this was not the prime reason for the death of so many companies. It took at least two generations before markets became saturated and even then the replacement market would have been sufficient to keep more manufacturing firms in business. The prime cause was the increasing efficiency and increasing complexity of manufacturing methods, which enabled fewer firms to produce more cars. In other words, even with only a handful of manufacturers left, manufacturing capacity outstripped market demand. The increased complexity of the product and associated high costs of development in a competitive market did the rest toward forcing a high concentration of firms. Whereas early motor vehicles were somehow cobbled together by skilled and imaginative individuals equipped with simple tools and standard engineering components, a modern car is the product of thousands of hours of planning, research, design, and ordering of special production machinery and special components. It requires a large firm with huge resources to finance all this outlay and to muster the qualified workforce to pull it off. Whereas the requirements for qualifications on the shop floor have decreased, requirements at the planning and design stages have increased. Unless a certain popular model is sold in very large numbers, the costs of development will not be recouped. Only luxury cars, which sell at premium prices, can be marketed profitably in small numbers.

The concentration in the industry is truly remarkable. Even the United States have only three[3] sizeable motor manufacturers; France has two, Italy has one, Britain has one rather small one, Germany has three and Japan has three or four. Most of the firms now operate on a worldwide scale. When we say that Britain has only one small motor manufacturer, this means only that there are no indigenous British owned firms manufacturing cars on any significant scale; it does not mean that motorcars are not manufactured in Britain by global firms. Ford, General Motors, Honda, Toyota, Nissan and BMW all manufacture motorcars in Britain, but Rover is the only British firm still in the business.[4] Indeed counting national firms has become largely meaningless, as most motor manufacturers are now worldwide players with factories in many countries and with complex linkages between them.

The example of the motorcar illustrates several basic general properties of technology. First, specific preconditions must be fulfilled before a particular innovation can come to fruition. The preconditions for the development of the motorcar included the development of a whole range of ancillary technologies, required both for the development of the car and its engine, and also for a wider system of transport capable of using the motorcar. This brings us to the second general property of technology: many technologies form systems. Roads had to be improved and the road network extended. When cars were first introduced onto the roads built for horse-drawn vehicles, dust and mud became major problems. All the early cars, up to about 1920, were open and their occupants had to wear goggles, leather helmets and dust-suits. As cars spread, methods of road building and surfacing were improved, e.g. by covering them with tarmac. However, as late as 1913 roads in Michigan were so deplorable that Ford could not deliver cars by road to customers living more than 100 miles from its Detroit factory.

---

3 All three US carmakers are global companies based in the USA.

4 By the time of revising this chapter in early 2009, even Rover has disappeared from the scene.

The roads invented and propagated by John McAdam in the late 18th and early 19th century consisted of a compacted and drained substrate, topped with a layer of large stones, followed by a layer of smaller stones and finished with a layer of gravel. The method was not that different from Roman methods, except that grading and transport of stones had become easier and steamrollers improved the compacting beyond recognition. McAdam roads served well for horse-drawn traffic, though even then they became alternately slimy with wet horse manure or dusty. Experiments with asphalt (a mixture of bitumen and gravel) as a surfacing material for roads or footpaths go back to the beginning of the 19th century and progressed greatly in the hands of Edward de Smedt in the late 19th century. Apparently the incentive for asphalting Paris boulevards soon after 1848 was to remove cobblestones from the reach of rioters to use as missiles or as barricades. In 1877 Pennsylvania Avenue in Washington DC became one of the first roads to be covered with the new material in the United States. The combination of the McAdam method of building roads and covering them with asphalt proved the decisive step toward modern roads, the tar mcadam, or tarmac[5] road. Apart from the need for improved road surfaces, the spread of the motorcar was dependent on an expansion of the road network. This required major civil engineering works, including the construction of bridges, viaducts, and tunnels. One of the more fortunate circumstances of the introduction of the motorcar is the fact that bitumen is a by-product of refining crude oil into petrol (gasoline). Thus crude oil serves the dual role of providing the fuel that propels cars, as well as the substrate that cars travel on.

Another major component of the motor transport system is fuel. The system consists of the exploration and drilling for oil, the setting up of a gigantic petrochemical industry which converts crude oil into motor fuel and a variety of other chemical substances, and a complex distribution system, including a dense network of filling stations that supply the millions of motor vehicles with their lifeblood. The spread of the motorcar was conditional upon the development of the network of fuel stations and vice versa. No point in buying a car if fuel is not widely available, no point in building fuel stations where there are no motor vehicles to use them. No doubt this mutual dependence slowed down the initial pace of introduction of motor vehicles. So important has the car become that the mastery over oil has become a central theme of power politics and of international conflict. Oil is, of course, also a major resource used in the manufacture of plastics and other chemicals.

The early motorcars could normally be repaired and serviced by any competent mechanical workshop, something like the latter day village blacksmith. As motor vehicles became more complex, it became necessary to establish a network of specialist car repair and servicing shops. As the number of manufacturers decreased and servicing became even more complex, with computer diagnostic devices and all that, many of the repair and service workshops became franchised and specialised for particular makes of vehicles.

Roads, fuel, workshops and dealers are by no means all that is required to operate a road transport system. We should not forget that the operation of road vehicles on public roads needs to be strictly regulated by a large body of law and regulation. Without regulation, traffic would deteriorate into utter chaos and come to a standstill. Laws and regulations need to be enforced and this means a huge apparatus of traffic police, traffic wardens, court cases, fines and so forth. As traffic is dangerous and any driver can cause or sustain very substantial damage to property or life and limb, motor vehicles have to be insured. All vehicles must carry a minimum insurance demanded by the law in order to guarantee that they are in a position to pay for any damage they might cause. Many drivers carry additional insurance to pay for damage that might be caused to their own vehicles or their occupants. As driving cars in modern traffic conditions requires a great deal of skill, a whole system of driving schools and, more importantly, examining and licensing drivers has become established. Despite all the regulations and all the driving tests, road traffic causes very large numbers of deaths and injuries. This means that a fleet of ambulances is required to ferry the victims to hospitals and hospital wards and their medical and nursing staff carry a large load of caring for victims of traffic accidents. Perhaps the word accident is a euphemism. Though undeniably some accidents do happen and may be unavoidable, the majority of so-called traffic accidents are caused by careless, unscrupulous, and aggressive use of the great power that the car puts at the effortless disposal of its driver. The car in the hands of a thoughtless driver is a potent weapon. In the early days the motorcar set out to capture the road space in more remote parts of cities and deprive children of its use as

---

[5] Tar is often used synonymously with bitumen

a playground. The result was carnage among children: by 1910 motorcars had killed over a thousand children in New York.[6] But still the motorcar marched on.

The motorcar is the prime example of the truly mass-produced complex product. This happened in several stages. The first stage was the production of standard components made with great precision. We have seen this happen with muskets first and the precision-made standard component became the foundation of mass production in mechanical engineering. To achieve the required precision, it was necessary to introduce accurate measuring techniques and, more importantly, accurate machine tools. It was also necessary to introduce standard screw threads and standard specifications for steel. All these preconditions had to be fulfilled before it became possible to assemble a final product from a stock of suitable interchangeable components.

Henry Ford, not content with the possibilities of simply assembling automobiles from standard parts, took the next steps toward even more efficient mass production. In today's parlance it would have been called mega production. He broke the assembly process down into very small steps and let each worker perform only a single one of these steps. Thus the division of labour was taken as far as was reasonably possible, with each worker carrying out only a small step toward the complete product. In order to speed up the process even further, the product was put on a conveyor belt that would take it past the workers, thus eliminating any wasted time between the individual steps in the production process. Each worker was supplied with the components he (or, later, also she) was to add to the assembly and was provided with the tools required for this particular task. The pace of work was dictated by the speed of the conveyor, the leeway and discretion of each worker was reduced to nothing. The workforce itself becomes a kind of machine with each part performing its allocated task. The transfer of power from workman – the blue collar worker – to engineer in the office – the white-collar worker – is complete. All flexibility, all initiative, all decision making and autonomy, and most skills are taken away from the worker and are incorporated into an engineering blueprint and detailed instructions for a small task. The assembly line was born and is alive to this day, though many operations are now performed by mechanical robots instead of by human ones.

To produce cars even more economically, Henry Ford standardised not only the working processes but also the product. The model T Ford that came on the American market in 1908, and remained in production till 1927, was sold in several body styles, but the chassis and the engine and everything else was completely standardised. More than 15 million model T cars were produced and its price was sufficiently low, and its maintenance sufficiently simple, to make it a real people's car. The model T enabled American farmers, who lived in very isolated small communities, to break out of their isolation and they bought the car in their thousands.

The Ford Company soon expanded its manufacture to Europe and European manufacturers soon introduced conveyor-belt methods to their factories. In France, mass production methods were pioneered by André-Gustave Citroën when he worked for the now defunct Mars automobile firm. In Britain, it was chiefly William Morris (later Lord Nuffield) who pioneered production methods closely similar to those of Henry Ford.

Between 1903 and 1908, before the model T and mass production methods were introduced, Ford cars cost around $1,600; a great deal more than the Oldsmobile produced by Ransom Eli Olds, that sold for between $400 to $500. By 1908 there were 24 companies building cheap cars, but Ford was not among them and continued producing cars for the wealthier customer. But soon things were to change. Henry Ford became interested in three things: vanadium steel for the production of valves; assembly line production; and so-called scientific management preached by Frederick W. Taylor Though none of these were Ford's original ideas, he applied them to entirely new problems and introduced them with greater thoroughness and greater success than anybody before him. Vanadium steel was a technical improvement and made his engines more reliable, but did not bring about a major change. The use of a production line for motorcar manufacture was revolutionary and was one of the important steps that changed the world. Henry Ford changed the perception of the purpose of motorcars. Instead of viewing the car as a toy for the rich, he looked mainly at isolated rural communities and felt that the car would help them break out of their isolation. By producing a cheap car he envisaged mass ownership and sought his profits in large numbers of cars, each sold at a modest profit instead of small numbers sold at large

[6] Ruth Brandon, (2002), p. 49

profits. His formula worked out for him – he became the richest man in America – and it produced, for better or for worse, the mass ownership Ford had envisaged. Perhaps no other technology has ever had such profound effects upon society and upon individual lives.

The first Ford car to be specifically designed for mass production and built by the new methods was the revolutionary T-model. Shortly before the introduction of the T-model, Ford raised the wages of his workers and thus raised their morale and productivity. The cheapest of the three variants of the T-model cost $440; half as much as the next cheapest decent car on the market. Car production and ownership increased by leaps and bounds. The Ford workforce in the main Detroit plant – the River Rouge complex – reached 42,000 by 1924 and over 100,000 by 1929. Whereas in 1898 there was one car for every 18,000 people, by 1923 there was one car for every eight people in the USA; 13million cars altogether.

Ford believed in total control and total ownership of production. He manufactured his own steel and even went as far as owning rubber plantations. This method obviously worked for him, but was not emulated by all manufacturers and has since been almost entirely reversed. Modern car manufacturers have, in the main, become assemblers of parts produced elsewhere. For about 12 years all model T cars were black and Henry Ford is reputed to have said: "the customer can choose any colour as long as it is black". The reason for this was technical: black enamel was the only fast-drying enamel available. The situation changed when Du-Pont brought out fast drying enamels in different colours in 1924. In 1919 87% of American cars were open; in 1931 92% were closed. The change was brought about by improved methods of manufacture of pressed steel components. A closed version of the model T became available in 1925.

Financial measures contributed to the mass ownership of cars. By 1925 about 65% of cars sold in USA were paid for in instalments and by 1927 about half of the new cars sold were replacements for cars traded in as part payment.

In 1927 the model T Ford was replaced by the model A. This included many technological improvements, such as a self-starter, a new gearbox for easier gearshift, hydraulic shock absorbers, larger pneumatic tyres, a safety-glass windscreen, and rubber-insulated seat cushioning. Acceleration was much faster and the model A could maintain a cruising speed of 50mph. It sold for $495, about $100 cheaper than the rival Chevrolet. The introduction of the self-starter was a major contributing factor to the fact that more women became drivers. Thus the car did help the farmer's wife as well as the farmer to break out of their rural isolation. The starting handle required quite a lot of strength and starting early cars was an altogether tricky manoeuvre. The car had matured into an artefact that required little skill to be used and transported people with reasonable comfort at reasonable speed. A large number of further improvements separate the model A Ford from the car of the present, but no further change in fundamentals has taken place.

Once the mass-ownership of cars had become established, manufacturers became anxious to retain mass sales despite market saturation. In the immediate post-depression years General Motors, headed by Alfred Sloan, succeeded in making the car an item of fashion. A new model was produced each year, just sufficiently different from last year's model to make it desirable. A new breed of designers, mostly concerned with the external appearance of the car, gained great influence. Whereas the car had been a pure engineering product, it now became a product of the stylist as much as of the engineer. The car had developed from a symbol of wealth and a toy of the rich into a symbol of affluence and relative wealth for the not so affluent. The choice of car is influenced as much by its appearance and its image as by its practicality. Owning a sporty car reflects a sporty image upon its owner, owning the newest and most expensive car in the street bestows an aura of wealth. Owning an old, cheap, second hand car shows either the poverty or the frugality of the owner. The designer in the late thirties and the first post-second world war decades created phantasies and dreams in their attempts to sell more cars. It took the courage of Ralph Nader with his 1965 book "Unsafe at any Speed", and a generally more sceptical attitude, to bring cars at least partly back to earth and make them a little safer. The dream-car was shown to be a nightmare-car that killed its naïve owners quite unnecessarily. Governments reluctantly accepted their obligations as guardians of the public interest and introduced regulations forcing manufacturers to introduce certain safety features. Eventually safety became a selling point and at least some customers began to apply safety as a criterion for their choice of car. Although cars are now very much safer than in earlier times, they still perpetuate many follies of previous generations. Speed is still stressed and all cars can exceed legally permitted maximum speeds by huge margins. Acceleration and sporty image are still stressed by sales departments and many a modern car is more akin to a projectile than to a means of transport. The huge popularity of motor racing undoubtedly contributes to making customers susceptible to the lure of speed.

The mass ownership of cars has had the most far-reaching consequences. Congestion of cities has spread from their centres into the whole of the city. Though cars have cleansed city streets from horse droppings and the stench of urine, as was hopefully expected of them, they have substituted pollution of a more insipid and more dangerous kind. The pollution from car exhausts is one of the major factors in the much-feared greenhouse effect and many ills, from corrosion of ancient buildings to asthma, have been attributed to exhausts from motor vehicles. The car has undoubtedly contributed to the sprawl of urban areas and to the decay of many central areas. The car has caused the rise of vast shopping centres and the death of numerous small shops.

The tractor, a close relation of the car, has replaced the horse as a farm animal and thus freed much land from its use for growing animal fodder. First came the horseless carriage and the horseless plough soon followed. By 1929 there were about 826,000 tractors working on US farms and the phenomenon of overproduction of food for human consumption reared it ugly head. Whereas in cities the small shopkeeper disappeared, it was the small farmer who disappeared from the countryside. It may be argued that all human progress demands change and adaptation, but the tremendous suffering of individual families should not be overlooked. Indeed pushing change too rapidly increases the sum total of human suffering rather than the sum total of human happiness.

The production of internal combustion engines was never as simple as building early cars and was never as widely distributed among small manufacturers. Most small makers of cars bought in engines from more specialised firms. Apart from the fact that casting and machining engine blocks is no trivial matter, the design of engines required a good deal of technical and scientific knowledge; indeed it was one of the early instances when engineering became based upon scientific theory. The steam engine required some scientific knowledge; the internal combustion engine required much more and contributed to the border between science and engineering becoming rather fuzzy.

The concentration process in the motor industry is truly remarkable and demonstrates a general property of technology. When a new product becomes highly sophisticated, and its manufacture becomes both complex and specialised, most early entrants into the new field are forced to drop out. They either disappear completely or merge with other firms, so that after some time only a few large firms are left through a process of selection or by coagulation. The automobile industry in other countries fared much the same as in the USA. Some of the earliest entrants are still around, but not many. In France Louis Renault built his first automobile in 1898 and the Renault firm is still a major manufacturer. The experienced automobile engineer André Gustave Citroën was engaged in the production of munitions during World War I and formed his own car manufacturing firm Citroën Cars, in 1919. His first and many of the firm's later cars were technically highly innovative models. The firm, though merged with Peugeot since 1976, is still very much alive within Peugeot-Citroën, known as Peugeot SA since 1979. The Peugeot automobile firm, founded in 1890 by Armand Peugeot, arose out of a small family-owned workshop producing velocipedes and quadricycles. Many others have gone under, among them Delage, Delahaye, Talbot, Voisin, De Dion-Bouton, Panhard-Levassor, and Simca.

British car manufacturers fared worse than their American counterparts. The largest early British manufacturers were Morris and Austin. The Austin Motor Company was founded by Herbert Austin, later Baron Austin, in 1906. The Austin Seven model was perhaps the nearest European equivalent to the Ford T and the firm remained at the forefront of European design until its merger with Morris to form the British Motor Corporation. William Richard Morris, later Viscount Nuffield, started his entrepreneurial career with a small bicycle repair shop that also built bicycles to order. Later he also repaired motorcycles, but he and his partner went bankrupt in 1904. As all true entrepreneurs he did not give up but set up shop in Oxford and produced his first car, the Morris Oxford, in 1913. His firm prospered and went from strength to strength. Morris Motors Ltd., founded in 1919, introduced Ford production methods and expanded by various acquisitions. One of the firm's most successful small cars was the Morris Minor, designed by Alec Issigonis, which came on the market in 1948 and remained in production till 1971. The successful design of technically advanced small cars continued in the

British Motor Corporation with its most famous revolutionary small car, the Mini, now produced by BMW in Britain. Most other famous British names, such as Jaguar, Lotus, Rolls Royce and Bentley, are now in foreign ownership and many others, such as Wolseley, Lanchester, and Hillman have disappeared. On the other hand, many Japanese companies have set up shop in Britain and some US-owned global companies contribute to an active motor industry in Britain, though almost none of it is in British ownership.

The German motor industry fared rather better. Volkswagen, originally a firm founded by the Nazi regime for the production of a "people's car" (Volkswagen) was almost totally destroyed during World War II. It was revived in post-war Germany with government aid and later became a private stock company, though with a share of public ownership, and one of the largest and most successful motor manufacturers in the world. Opel, properly Adam Opel A.G., was started in the earliest days of the motorcar when in 1898 the five Opel brothers began converting the Adam Opel bicycle and sewing machine factory to motorcar production. Their first successful car came on the market in 1902 and from that date Opel went from strength to strength until 1911, when fire destroyed the factory. Opel rebuilt and used the opportunity to bring the factory up to the most modern standards. After World War I they introduced Ford methods of production. Despite this, the Opel family sold out in the 1920s to General Motors and Opel has been part of GM ever since. BMW is the most recent of German car manufacturers. Founded in 1929, it produced motorcycles and aero engines. Cars were added to its programme after World War II and became successful in the 1970s. The oldest of German carmakers is Daimler Benz, discussed earlier.

The story of the many mergers in the motor industry is a symptom of three phenomena. First, the motor industry, like many other industries, now operates on a global scale. Production is spread over many countries and marketing is worldwide. Secondly, that concentration into fewer and fewer gigantic firms is continuing, though it might possibly have reached some sort of stable state. Some observers argue that further concentration is being replaced by cooperation between rival firms on individual projects. Joint ventures between rivals may be a substitute for a further reduction in the number of independent manufacturers. A third phenomenon is the mutual share ownership. It is not uncommon for one large manufacturer to own shares in another and such shareholding may be reciprocal. The total number of global players is now a mere handful.

In the early days of the automobile it was not at all obvious which means of propulsion should be used. There were three contenders: steam engines, a combination of battery and electric motor, and the internal combustion engine. Eventually the internal combustion engine got ahead, mainly on grounds of weight and size. The battery simply was not up to the task of carrying sufficient energy at a reasonable weight and the steam engine suffered from the disadvantage that it needed a warming up period before the car was ready to start. It should be stressed, however, that so-called flash boilers could overcome this problem and that the steam engine has the advantage of not needing a gearbox as, unlike the internal combustion engine, it can develop very high torque at very low speed. The petrol engine got ahead because much more development effort was put into it, especially when the first mass producers, such as Ford, chose the petrol engine and developed the engine and the whole system associated with it. Once the petrol engine had made so much headway, the steam engine no longer had a chance of catching up. Too much had been invested into the development of the petrol engined car. At the turn of the century the outcome of the competition between steam and internal combustion was still undecided. In 1901-2 the best-selling car on the US market was a steamer and there were 50 manufacturers of steam cars around. The splendid luxurious and expensive Stanley steamer continued in production till 1927.

The development of the engine required a great deal of scientific and engineering knowledge. The choice of materials, the casting of the engine block, the choice of an operating cycle, the arrangements for adequate cooling, for ignition, for carburation and so forth were all quite tricky. Indeed the internal combustion engine marks the beginning of the era in which engineering and science became closely interwoven.

The initial stimulus for the development of the motorcar was twofold: first, the possibility was there for all to see because of the earlier development of the steam locomotive, and secondly it seemed obvious that wealthy people would love to play around with this attractive new toy. The promise of speed, the promise of control over considerable power, the possibilities of displaying wealth, manliness and modernity were temptations that ensured an adequate market for quite a few very small craft-based manufacturers.

Nobody in the early days did foresee the later enormous growth in ownership and the far-reaching consequences that the introduction of motorcars on a massive scale would have. The leap between producing cars in ones and twos, or even in hundreds, and producing millions of cars is so large that any social problems that might be caused by the car changed not simply in quantity but in quality. From very small and rather insignificant beginnings the motorcar changed into a potent mix of blessing and curse for humanity.

Whereas in the pioneering days of the motorcar one could identify a few outstanding individuals who advanced this technology, and in later years one could identify some entrepreneurs who advanced the industry, the modern enterprise is dominated by large anonymous bureaucracies, whether in engineering, in styling, in sales, or in administration. There are no more heroes – perhaps the last was Sir Alec Issigonis who became a hero when the Mini designed under his leadership achieved cult status.

We can do no more than provide a sketchy outline of the main social consequences of the introduction of the motorcar. To do full justice to the topic requires a separate book. Many have been written and some are listed in the bibliography.

The main advantage of the private motorcar is, of course, that it provides personal mobility including, for good measure, a substantial load-carrying facility. The car is supposed to get you and your luggage and your passengers from A to B at speed and in comfort. But the big cities have become so congested that travel in the car is neither fast nor comfortable. Though it protects the driver from inclement weather, it exposes him or her to a great deal of stress and frustration. And where do you park at either A or B?

The car and other motorised transport has become a major cause of air pollution with all its consequences, particularly the so-called greenhouse effect that threatens to cause highly unwelcome changes in the climate of the Earth. The car and other motorised transport have also become the major cause of urban sprawl and the geographic growth of cities.

The car has become a major cause of death and injury, as we seem unable to master the problem of road accidents. In some younger age groups, the car is the largest cause of death.

In conjunction with the refrigerator and the home freezer, the car has changed shopping habits. Car owners now shop but infrequently in large supermarkets that have displaced most small family-owned shops. The young, the disabled, the poor, and the elderly are excluded from driving and, because all life is now planned around the car, suffer severe disadvantages.

Motorised traffic has become a major consumer of energy, particularly oil. Oil is a non-renewable source of energy and our reserves are bound to run out, sooner or later. The oil industry is highly concentrated and extremely powerful and oil has become a major cause of international conflict.

Unfortunately the excessive use of the motorcar can only be curbed by shifting public investment from roads to public transport and by some suitable regulatory measures. So far, governments have not found the will or the courage to act – except by way of lip service - in the face of a strong combined lobby of the motor and allied industries and the motoring organisations. Despite its severe drawbacks, the car is still the beloved child of citizens and governments. However, using the car as a means of mass transportation is in irreconcilable conflict with its very nature as a vehicle for personal mobility. It is not enough to legislate for more frugal and safer cars, it is also necessary to find means of confining the car to use for individual transportation in situations where this can be done without too much interference with other users, and to use systems of mass transportation for mass transportation.

Aviation

We take the development of aviation as our next example of the enormous changes that the twentieth century brought about in conquering, in a sense eliminating, distance. Whereas the pedestrian can cover, say, 40km in a day; the rider or the horse-drawn carriage can cover, say, 120km. If we allow for the possibility of travel at night, a modern car can cover something like 1,600 km in 24 hours. The modern passenger aircraft, cruising at a speed of about 900km/h, and needing a refuelling stop only after about 16,000km, can theoretically cover roughly 20,000km in 24 hours, i.e. halfway round the globe at the equator. Thus, very roughly, the distance horizon of the individual has increased by a factor of 500. Perhaps more importantly, travelling by plane it is possible to reach any point on earth within 24 hours, whereas travelling on foot or on horseback it took weeks or months to reach every point on any continent, let alone in intercontinental travel in sailing ships. Air travel and telecommunications have shrunk the planet from superhuman to human dimensions. These were factors, or preconditions, for the much praised and much cursed present day globalization. On the other hand, the traveller on foot causes virtually no damage to the environment and only needs about one cubic metre of space to move in; whereas the modern aircraft needs huge amounts of fuel, produces very large quantities of highly damaging emissions, and occupies several cubic kilometres to fly at safe distances from other aircraft. We have become rather large and greedy monsters.

Modern air travel, like all technologies, had several precursors. The hot air balloon was invented by the brothers Montgolfier as early as 1783. It was constructed from paper and before the year was out, Jean-François Pilâtre de Rozier and François Laurent, Marquis d'Arlandes made the first manned flight in a balloon. Perhaps the most important and earliest pioneer of flight with heavier-than-air machines was Sir George Cayley. After some experimentation with model gliders he worked out the aerodynamic theory of aircraft wings in 1809, as well as the general layout of planes with a fuselage, wings, and tailplane with rudder and elevator. It was Cayley who designed and built the first glider to carry out a manned flight in 1853. The line of scientific predecessors to flight is even longer, for Cayley based his theory on the theory of fluid flow, established in 1738 by the Swiss mathematician Daniel Bernoulli, and on experiments by the Italian physicist Giovanni Battista Venturi. In essence, the Bernoulli theorem is a form of the law of conservation of energy. When the velocity of flow of a fluid increases because of a constriction (e.g. in a venturi tube), the pressure decreases. It follows that if air flows past a correctly designed wing, the pressure below the wing will be greater than the pressure above, thus providing lift. In the second half of the 19th century Otto Lilienthal and his followers made thousands of flights in a variety of homemade gliders. By the end of the century several kinds of balloon had flown quite extensively. Gliders and balloons may be regarded as the forerunners of powered flight, though neither of these forerunners could be flown at will from A to B and cannot therefore be regarded as proper means of transport.

The first sustained, controlled powered flight, albeit over a very short distance, was carried out in December 1903 by the brothers Wilbur and Orville Wright. They were well aware of Cayley's theory and had themselves carried out many systematic experiments. They made use of a strong headwind to provide the lift that their feeble engine failed to provide. By 1905 the Wright brothers had made some progress and managed to make circular flights of up to 24miles. Their plane was a biplane with two airscrews driven by a chain from a petrol engine. The next important pioneer of powered flight was Louis Blériot, who crossed the English Channel in July 1909 in a monoplane with a forward mounted airscrew (so-called tractor monoplane) driven by a petrol engine. A successor to Blériot's early plane was produced in fairly large numbers for a growing band of flight enthusiasts. Another successful early plane was a so-called pusher biplane built by Gabriel and Charles Voisin. One of Voisin's planes was used by Henri Farman to improve controls and other facilities and by 1909 the Voisin pusher biplane had reached something like a stable design. Louis Breguet achieved the same for the tractor biplane at the same time. These two designs dominated the next two decades, though further developments of engines and improvements of aerodynamic stability of aeroplanes continued.

It became obvious that only advances in motor design and in aerodynamic theory could improve the performance of aeroplanes. In aerodynamic theory three early pioneers stand out: F.W. Lanchester, Ludwig Prandtl and Albert Betz were the founders of this new branch of applied physics. This is another example of the increasingly close relationship between theoretical physics and mathematics on the one hand and engineering on the other.

The First World War accelerated the development of aeroplanes. It has been estimated that a total of 5,000 aeroplanes had been built by 1914 and that this number had risen to 200,000 by 1918. In the early part of World War I planes were used almost exclusively for reconnaissance. Typically they were two-seaters with speeds of 70-80mph (115-130 km/h) and a range of about 100 miles (160 km) and could fly at a height of up to 10,000 feet (3,000 m). By 1916 many of these planes were equipped with machine guns firing between the blades of the propeller, and by 1917 a specialized fighter aircraft had been developed and air-fights became common. The fighter was usually a single-seater, equipped with two machine guns, flying at speeds between 160 and 200 km/h, with a ceiling of about 6,000 m. During the later stages of trench warfare, the role of aircraft was extended to bombing, with trenches a favourite target. The Allies used mainly bomber aircraft for the purpose; the Germans used both aircraft and airships.

Aero-engines provide an interesting example of measurable technological progress. One of the characteristic figures of performance for aero-engines is the output per unit of weight; obviously we wish to obtain the greatest possible power from the least possible weight. In 1880, Otto engines weighed about 200kg/hp. In 1890, an engine designed by Daimler and Maybach weighed only 30kg/hp. It is interesting to note that the power to weight ratio of a typical petrol engine of a contemporary ordinary family car is about 1kg/hp. It is hard to know how this compares to the Daimler engine because many things, such as the clutch, may be included or excluded from one or the other figure. In 1910 the popular French Antoinette aero-engine weighed 95kg and produced 50hp (1.9kg/hp) and the Gnome, designed by Laurent Senguin in about 1914 weighed only 75kg for 50hp. In 1918 an American Liberty water-cooled V12 engine weighed only 1kg/hp.

The era of civil air transport began after World War I and did not get fully into its stride till after World War II. In the early years after World War I it seemed that the airship, a kind of rigid powered and dirigible balloon, would play an important role. Airships were built in several countries, including Britain, the US and Germany. The name of a German pioneer of the airship, Graf (count) Ferdinand Zeppelin has become a synonym for airship. He achieved 24-hour flight in 1906 and, as a consequence, obtained an order for about 100 airships for the German army that used them during World War I. The rigid airship is not the only possible powered lighter-than-air flying machine. In 1852 H. Giffard attempted to power a flexible balloon with a steam engine and others have built flexible and semi-rigid airships. With hindsight, many ideas and experiments in the history of technology look rather bizarre. After Zeppelin's death his firm continued to build rigid airships. For reasons of cost and for political reasons these were filled with hydrogen, rather than with the much safer non-inflammable helium. The first non-stop transatlantic Zeppelin flight took place in 1926. The large passenger airship *Graf Zeppelin* flew round the world and with its sistership *Hindenburg* established a regular route from Germany to the USA. Airships are comfortable and spacious in good weather, but they are very slow compared to heavier-then-air aircraft and get buffeted by the wind because of their low altitude. They also present serious problems of handling on the ground and when the *Hindenburg* burnt out on landing in the USA in 1937, the use of airships was virtually discontinued. The occasional attempt is still made to revive the airship as a cargo carrier, but so far this has not come to much. Normal aircraft and helicopters appear to fulfil all the requirements.

The development of airships was heavily subsidised by the public purse and indeed all aircraft development was carried out mostly at public expense. The military bore the brunt of the development costs of military aircraft and the costs of civil aircraft development were greatly reduced by military pioneering work.

The first purpose-built airliners of the 1920s were biplanes constructed of wood and wire, very much in the Wright tradition. Toward the end of the decade they were made of steel and duralumin.[7] Hugo Junkers was the first to build a practical all-metal plane from corrugated duralumin sheet. The skin bore part of the load; the rest was borne by struts. His firm produced a single-engine all-metal low-wing monoplane passenger aircraft in 1919, the F-13. At roughly the same time the Dutch firm Fokker produced a high wing single engine monoplane with a welded steel tube fuselage and wooden wings. In 1925 Fokker produced the first practical multi-engined aircraft, the F.VII-3m, powered by three motors and built from corrugated-duralumin sheet. Junkers were more successful with their Ju 52/3m that became the standard three-engined plane for most European airlines in the late 1920s and the 1930s. American airliners began making inroads into European markets shortly before World War II.

In 1920 the British firm Handley Page produced a 12-seater passenger aircraft that also proved influential in civil aviation. The company had been founded in 1909 by Sir Frederick Handley Page and had produced the first twin-engined bomber during World War I. Another British aircraft pioneer, Sir Geoffrey De Havilland, founded his company in 1920 and became influential in amateur club flying by producing a light two-seater, the

7 Duralumin is a strong alloy of aluminium with small additions of copper, manganese and magnesium and sometimes other metals. It was originally patented by Alfred Wilm in Germany and now exists in many variants with properties adapted to particular tasks.

Moth. The company became famous for its production of the *Mosquito*, a small all-purpose military plane, constructed mainly of plywood, that played an important role during World War II. De Havilland also produced in 1952 the very first civil jet-engine airliner, the ill-fated Comet; temporarily withdrawn after 2 years when three fatal crashes revealed a flaw in the design: metal fatigue caused a sudden fracture of a vital part of the fuselage in mid-air. Metal fatigue[8] was at that time little understood and designers had to learn from this bitter experience how to cope with it. After a painstaking investigation and a careful redesign, the Comet went back into airline service for several years. The leadership in jet-propelled passenger transport was, however, lost to Britain and transferred to the USA.

The air-cooled radial piston aero engine became standard from the late twenties. Intervals between engine overhauls were gradually lengthened from about 150 to 200 hours in the early 20s, to 400-500 hours in the 30s. Despite this and despite expensive fares, airlines were unable to cover more than about 25 to 30% of their costs and depended on government subsidies. Among the early European airlines were KLM, founded in 1920; Sabena founded in 1923; Imperial Airways (1924) and Lufthansa (1926). Some of these were formed by mergers from predecessors; all maintained their expanding networks with the aid of subsidies.

Airline services in the USA started a little later than in Europe, but by the mid-twenties expanded rapidly and by the end of the decade American airlines carried more passengers than their European counterparts.

Monocoque designs with stressed skin structures were introduced in the late twenties and early thirties and became the norm for airliners by the mid-thirties. Among the pioneers were C. Dornier, A. Rohrbach, G. Baatz and, in particular, the theoretician H. A. Wagner. Virtually all the pioneers of powered flight were academically trained engineers and progress was possible only with the elaboration of complex and accurate theories and systematic experimentation. The all-metal transport plane became standard, as did the wing without external struts, improved aerodynamic design and increased size of multi-engined aircraft.

Other important improvements were introduced in those years. Retractable landing gear greatly improved the aerodynamic performance, wing sections were improved by better understanding of aerodymamic theory, flaps to improve lift on takeoff and landing were introduced and the engines became fully cowled. Variable pitch propellers were another innovation and instrumentation expanded and improved. Operating techniques of airlines and airports improved and thus the whole business of passenger transport by air reached some form of preliminary maturity. Apart from aircraft made by Junkers, Fokker, Handley Page and other European manufacturers, the American Boeing 247 and Douglas DC2 and its successor, the DC3 (in 1935) became standard equipment of the airlines.

Unhappily much aircraft development was aimed at war. The standard bombers of World War II were four-engined planes capable of carrying large loads of deadly cargo. Fighter aircraft were also produced in very large numbers and used mainly to attack or defend bombers. The numbers produced were staggering. The US alone produced over 700,000 military aircraft during the war period. The role of aircraft in World War II, both from the strategic and the tactical points of view, would be hard to exaggerate. Some bombing raids on cities have become part of the standard repertoire of history – the almost total destruction of Coventry in England and Dresden in Germany are just two examples of the horrors of strategic bombing. The controversy on whether strategic bombing was necessary and whether it much influenced the course of the war is still a subject of debate and will probably never be unanimously resolved. Most people believe, however, that when the Germans began to lose more bombers over Britain than their factories could replace, and thus the Battle of Britain was won by the British Spitfires and Hurricanes, this represented a turning point in the war that eventually led to the allied victory.

All improvements in aircraft design from the late 20s and early 30s to this day have been the result of enormous R&D and design effort. Table 6.1 shows a few estimates for the man-hours of design engineering time invested into the various aircraft before their maiden flights. (Source: Williams, History of Technology, vol VI, fig.33.18)

[8] Metal fatigue is the result of repeated stress reversal that leads, after many cycles, first to cracks on the surface of the stressed part and eventually to fracture.

Aircraft	Year	Man-hours
Spitfire	1936	300,000
DC3	1936	150,000
DC4	1942	1,000,000
Comet	1952	2,000,000
DC8	1958	10,000,000
Boeing 747	1968	30,000,000

Table 6.1 Man-hours of engineering design invested in various aircraft

These figures illustrate a fairly general law of technological development. As a technology matures, it becomes increasingly difficult to achieve further improvements. In other words, there are decreasing returns on invested effort. Any small improvement needs enormous effort to achieve. We call this fact the law of diminishing returns. Part of the design effort goes into ensuring lower operating costs. Indeed the direct operating costs of modern airliners per seat mile are considerably lower than the operating costs of earlier generations of airliners. Cruising speeds increased considerably in the early days of development, but have now stabilized at speeds just below the speed of sound. There was one single exception to this rule: Concorde. This Anglo-French airliner was built in very small numbers, flew at about twice the speed of sound, and its life as a commercial airliner finally ended in October, 2003.

Concorde[9] is not so much a civil airliner as a symbol of combined British and French national pride and engineering prowess. It is also an example of misguided government technology policy. Some aeronautical engineers at the British Royal Aircraft Establishment began looking at the feasibility of emulating supersonic military aircraft and building a civil version. As usual, a committee including aircraft industry interests and government was established and, in 1959, the committee recommended that a medium-range and a long-range supersonic airliner should be built by British industry. The estimated design costs were £50-70 million for the medium range and £90 million for the long-range plane. The idea for the medium-range design was soon dropped, but design studies for the long-range airliner began in earnest and were completed in 1961. The French manufacturer Sud Aviation announced its intention to build a SST (supersonic transport) at about the time of completion of the design study. It was also the time when the UK was beginning to show interest in joining the European Economic Community. The British and French governments agreed that talks on cooperation should begin between Sud Aviation and the leading British contender for the SST, the British Aircraft Corporation (BAC). After a year of discussions, a joint outline design emerged in September 1962. At this time the two-version plan had re-emerged and in November 1962 the two governments signed an agreement for sharing the cost of development. Britain's share was estimated at £75-80 million.

There was no demand from the airlines for such an aircraft; it was a clear case of so-called technology push. The motivation of the technologists was to attempt what had never been attempted before and thus showing their prowess. The governments involved went into it to show willingness for cooperation and to further the national prestige of French and British industry, especially vis-à-vis the United States. It was not a commercially inspired enterprise- the inspiration was technological and political. Having said that, the proponents of the scheme did try to make out a commercial case for it and produced wildly optimistic estimates for potential sales of 300 to 400 aircraft.

The medium-range aircraft was dropped again and detailed negotiations between Sud Aviation and BAC produced an agreed design in March 1964. On paper at least, Concorde was born that would be able to carry 118 passengers across the Atlantic. An existing Bristol Siddeley Olympus jet engine would be modified to produce a thrust of 32,825 lb. The project was given some urgency when American manufacturers began design studies for their own "bigger and better" SST. The commercial defeat of the Comet and VC-10 airliners by the big American jets had not been forgotten. In 1970 the American project came to an abrupt halt. Congress refused to give it the required financial support. One of the factors that influenced this decision was a growing

[9] The story of Concorde is taken from E. Braun, D. Collingridge, K. Hinton, (1979), Assessment of Technological Decisions-Case Studies, London, Butterworths

anti SST lobby that objected to the sonic boom that any aircraft creates when it goes through the sound barrier, i.e. when it exceeds the speed of sound. The objections had become very strong and it seemed unlikely that any government would permit the SST to fly at supersonic speeds over populated areas. By then environmental awareness had increased and people were worried about high fuel consumption of SSTs and damage to the ozone layer by aircraft flying at very high altitudes. Realistically, the death-knell for the project had been sounded, but those who should have listened were deaf to it.

The Anglo-French project soldiered on. The size of the aircraft was slightly increased and sales forecasts were, rather arbitrarily, increased to 500. Reality soon overtook the project. The project was completed and a technically brilliant aircraft was built and went into service in the early eighties – twenty-five years after its inception. A total of nine aircraft were sold at subsidized prices to two airlines: the national carriers of Britain and France, Air France and British Airways. The total development cost exceeded the estimate by a factor of more than ten and is thought to have been over £1,000 million. Flights in Concorde were very expensive – more than normal first class fares – because operating costs were high. Only very rich people in a great hurry used and loved this super-luxury service. Ordinary mortals fly in jumbo-jets.

After this supersonic detour we return to normal air traffic. In the immediate post World War II period the main workhorses of the airlines were former military transports, e.g. the Douglas DC3 carrying 20-30 passengers and the DC4 carrying 40-60 passengers. The DC3 was a short haul twin-engined aircraft, the DC4 was four-engined and built for longer routes. The cruising speed of these aircraft was only a little over 300km/h (200mph) and they had to fly at low altitudes because they were not pressurized.

A big step forward in long-haul airliners was made in the mid- to late fifties. Aircraft such as the Douglas DC-6 and DC-7, the Boeing Stratocruiser and the Lockheed Constellation, came into service. The successful short-haul Vickers Viscount belongs to the same period. Cruising speeds had increased to 480 – 530 km/h, the range had increased and all these aircraft were pressurized and flew at heights of about 5,000 to 6,000m, carrying about 100 passengers. So successful and popular had flying become that in 1957 more passengers crossed the Atlantic by air than by ship. This was an important crossover point and it was not many years later that regular transatlantic passenger shipping services were withdrawn altogether.

This generation of large passenger airliners was the last to use piston engines for propulsion. The piston engine was replaced by the gas turbine that operates on an entirely different principle. The air taken in is first compressed and then, in a combustion chamber, mixed with injected fuel. The expansion caused by combustion drives a turbine and this can either be used to drive a propeller – the so-called turbo-prop engine – or be expelled as a fast stream of gas that drives the aircraft by the force of reaction without further moving parts. We then speak of the turbo-jet engine. A variant of this engine uses a large fan for the air intake and allows some unmixed air to go directly into the jet, thus bypassing the turbine proper. Such engines are usually referred to as turbo-fan jets. The term jet engines covers both turbo-fan and turbo-jet engines because both propel the aircraft by a stream of gas rather than by a propeller. The propeller becomes inefficient at high speeds and generally the turbine has a higher power-to-weight ratio and is more efficient, as well as more reliable, compared to the piston-engine in aircraft applications.

The water turbine and steam turbine have been known for some considerable time and various designs of gas turbine have been proposed from time to time. In 1926 A. A. Griffith evolved a new theory of turbine blade design and in 1929 he proposed the turbo-prop engine using an axial compressor, i.e. a kind of reverse turbine to compress the air before combustion. Axial flow compressors are in universal use for all but the smallest gas turbines. A turbo-jet engine with an axial compressor was built by Metropolitan Vickers and powered a pure jet fighter aircraft that flew in 1943.

Best known as the inventor of jet propulsion in the English-speaking world is Sir Frank Whittle. He first proposed jet propulsion with a centrifugal compressor in a thesis written in the RAF College in 1928, and took out a patent on such a jet engine in 1930. Having failed to gain support from the RAF, he founded a firm Power Jets Ltd. In 1936 he built his first jet engine and tested it on the ground in 1937. With the beginning of the war the government became interested in Whittle and supported his work so that in May 1941 a Gloster fighter, equipped with a jet engine, flew on its maiden flight. The jet powered Gloster Meteor fighter soon entered RAF service and saw action during the later stages of the war.

In an independent development in Germany, Hans Pabst von Ohain, a physicist working for the aircraft manufacturer Heinkel, developed a very similar engine that was bench-tested in 1937. It first flew in the He-178 in August 1939 and an improved version was produced in 1941. The first operational German jet fighter, the Me-262 was, however, powered by a Junkers jet engine with an axial compressor developed by Anselm Franz. This jet fighter flew in 1942 and, like the British Gloster Meteor, entered active service in 1944.

We return to the post-war era and the 1950s and 1960s, when the turbo-prop and the pure jet entered service in civil airliners. Among the turbo-prop airliners the Vickers Viscount (1953) and the much larger Bristol Britannia, as well as the Lockheed Electra deserve mention. The turbo-prop era was short lived, although smaller turbo-prop aircraft are still in production and serve in various niche markets. The main airliner markets were soon captured first by the turbo-jet and then by the fan-jet.

Numerous contenders entered the jetliner market with many models of short-haul and several models of large long-haul aircraft. Standard cruising speed came near the speed of sound, 800 – 900 km/h, standard cruising altitude became 10,000 – 13,000 m and the various models of the 60s and 70s carried between 100 and 200 passengers. The development costs of the larger aircraft of the late 50s and early 60s had risen to about $300million apiece. The main contenders among the short-haul aircraft were the DC-9, Boeing 737, BAC 1-11, the BAC Trident, and the Sud Aviation Caravelle. The long haul was soon dominated by the Boeing 707 and the DC –8, although the Vickers VC-10, and the Lockheed Tristar managed to stay in the market for several years. It soon became obvious that no European manufacturer could compete with the two remaining American manufacturers: Boeing with their 707, 727 and 737 airliners and McDonnell Douglas with their DC8 and DC-9. Even Lockheed had to give up civil airliner production when the long-haul market became dominated by the jumbo-jet, the Boeing 747 and, to a much lesser extent, the McDonnell Douglas DC-10.

When it became obvious that individual European manufacturers could not survive against American competition, a consortium was formed in 1970 that became known as Airbus Industrie. Members of Airbus are Aérospatiale of France, Deutsche Airbus GmbH (fully owned by Messerschmitt-Bölkow-Blohm GmbH), each owning 37.9%; British Aerospace PLC, (since 1979) owning 20%; and Construcciones Aeronauticas SA (CASA) of Spain (since 1971) owning 4.2%. Several smaller companies participate as associates. The headquarters and main assembly plant is in Toulouse and parts are ferried from many locations, including major wing sections from Britain.

The first ever product of Airbus Industrie was the A300 short to medium-range twin-engined airliner that entered service in 1974. It was joined by a smaller medium-range plane, the A310 in 1978 and by the highly successful short/medium-haul A320 in 1987. The Airbus A-320 was first delivered to airlines in March 1988 and within ten years had accumulated orders for over 900 aircraft. It has a wingspan of 33.9m, is powered by two turbo-fan jet engines, delivering 118kN each, cruises at about 900km/h, and has a range of 5,400km carrying a typical payload. It accommodates 150 passengers in two classes, with a maximum of 179 passengers in a single class layout. There is a stretched version, the A-321 available. This has a slightly larger wingspan and slightly more thrust and can carry up to 200 passengers, and has a range of 4,260km with a typical payload. The largest Airbus model in actual service at the time of writing is the four-engined A340 with a seating capacity of up to 440 passengers and a range of 13,500km with a typical payload.

The Boeing 747, popularly known as the jumbo-jet, represents a major milestone in aircraft design and operation. It was hugely bigger than all its predecessors and it remained the dominant long-haul aircraft for the best part of 20 years. The 747 airliner was launched in 1966 and is still in production, though many technical changes have been made and many variants are available. A typical jumbo is the 747-400 that obtained its air-worthiness certificate in early 1989. The 747-400 is obtainable with a choice of several engines, at least one from each of the remaining three major aero-engine manufacturers (not counting combinations and cooperations), i.e. General Electric (USA), Pratt & Whitney (USA) and Rolls-Royce (UK). A typical engine is one of the Rolls Royce RB 211 series with 258 kN [10](58,000lb) thrust, giving a total thrust from the four engines of over 1,000kN. The plane has a wingspan of 66.44m, a total length of 70.66m, a height of 19.41m and a wing area of 524.9m^2. This

[10] KN stands for thousand Newton. Newton is the standard unit of force, defined as the force that accelerates a mass of one kilogram by one metre per second per second. 1kN is approximately 225lb.

gigantic plane can carry a maximum payload of 62,690kg and has a range of 13,180 km carrying the maximum payload. In a typical three-class layout it carries 421 passengers. The maximum cruising speed is 938km/h.

The latest and largest Airbus plane, the A380, designed to out-jumbo the jumbo, has been built and introduced to the public in early 2005. It will be manufactured in a new assembly plant in Hamburg. It is a truly enormous plane, it has two decks and two aisles and a range of 8000nautical miles (15000km). In a three class version it seats 550 passengers, in a single class it is said to be able to seat 700. It has two engine options: the Rolls Royce Trent 900 engine of 70,000lb thrust and an equivalent GEC/Pratt&Whitney engine. Many novel materials are used in the construction of this plane, such as carbon fibre reinforced plastics and a material consisting of layers of aluminium and fibreglass, known as Glare. The plane was launched in December 2000 and entered service early in 2006. There were considerable delays to early deliveries because of technical problems, but the plane is selling reasonably well and its numbers in service are growing steadily.

The largest cargo plane in the world at the time of writing is the Ukrainian Antonov AN-225. It has a wingspan of 88.4m and an overall length of 84m. It can carry a payload of up to 250,000kg, and has six turbofan engines of 230kN each. It cruises at 850km/h and has a range of only 2,500km with its maximum payload. This aircraft is designed to carry a space vehicle on its back from Moscow to Baikanour Cosmodrome. Antonov make a range of somewhat smaller freighters that are used throughout the world for carrying exceptional loads.

The commercial aircraft industry has come a long way since the early days of the Junkers corrugated flying boxes. Aircraft have become very much faster, very much larger, very much more comfortable, have a very much greater range and their operating costs per passenger mile (or kilometre) have become very much lower. All this follows the normal pattern of technological development: the quality and performance of the technology improves over time and it generally becomes easier to use. In the case of airliners this is very much the case. Navigation, aided by satellite positioning and other aids is now very simple, instrumentation and controls are all computer-aided, and autopilots have become highly effective. In the early days of transatlantic airliners there was a crew of four on the flight deck: two pilots, one engineer and one navigator. With advances in navigation aids the navigator was the first to go. With advances in aircraft controls the engineer disappeared next and now even the largest airliners have a crew of only two pilots on the flight deck Maintenance intervals have been increased and all functions are monitored by computer and faults are accurately diagnosed by computer.

What is also characteristic of technological development is the reduction in the number of competing manufacturers. After the merger between MacDonnell-Douglas and Boeing, the world aircraft industry producing large civil airliners has now shrunk to three firms. Boeing and Airbus in the Western world, and the Russian and Ukrainian industry in what used to be the Soviet world, are the only firms left. There still are several manufacturers who produce small aircraft, from the hobby-flyer to the commuter plane to the executive jet, and there are a small handful of firms that specialise in producing fighter aircraft. Concentration can hardly go much further. The main factor that has caused this extreme concentration is the complexity of modern airliners and the consequential extremely high development costs. The manufacturer needs sufficient sales to recoup development costs and this need reduces the number of viable manufacturers. The same has happened to engine manufacturers, whose number is now very small indeed, particularly in the field of large engines. The one factor that somewhat eases the situation of the aerospace industry, as it now prefers to be called, is the enormous amount of money spent by governments, especially the US Government, on defence procurement and development. But even there cooperation is the name of the game as single firms and even single governments often cannot afford the enormous development costs. On the other hand, each of the large firms buys in many components and sub-systems from specialist outside suppliers. Thus the number of firms in the aerospace industry is still substantial, but the number of suppliers of complete large aircraft is very small.

One feature of aircraft development is remarkable, though not unique: civil airliners (unlike their military counterparts) have stopped short of increasing their speed beyond the speed of sound. The speed of sound, also known as Mach 1[11], poses a real physical barrier. To go faster, everything needs to change. The shape of the most

[11] The Mach number is the ratio of the velocity of an object in a fluid to the velocity of sound in the same fluid. The velocity of sound in a gas depends upon its nature and upon the temperature and the pressure. In dry air at atmospheric pressure and 0°C the velocity of sound is approximately 330m/sec, equal to approximately 1,200km/h.

effective wing for supersonic flight is different from that for subsonic flight, the air intake for the engine needs to be different, problems of heating of the fuselage become severe, fuel consumption rises rapidly and, as soon as the aircraft flies at supersonic speed, it produces a loud so-called sonic boom. As mentioned earlier, the only supersonic civil airliner was developed for political rather than commercial reasons and the commercial industry has shied away from the problems of supersonic flight. It may come one day – who knows – but at the moment the costs and the problems are too great to justify the relatively small advantages to be gained. In fact this feature of technological development is to be found quite often. After a period of rapid development, certain features of the technology reach a stable state when very little further change occurs. Mostly this happens because a natural barrier makes further development difficult or impossible; sometimes it happens because of social barriers, sometimes simply because no worthwhile gain is expected from further development.

The speed of motorcars is an example of social barriers being hit. Most countries impose speed limits that are well below the design speeds of cars. Even when speed restrictions are largely absent, at least on motorways, as is the case in Germany, the tolerance of the public for extreme speeds is limited.

Before concluding our brief discussion of aircraft, we shall look at the specifications of a few modern fighting machines and, as in civil aviation, stand in wonder over the enormous development that has taken place since the earliest days of aviation, roughly speaking in one century.

As a first example we take the American B52 bomber made by Boeing, the oldest and one of the most awesome warplanes in the awesome US arsenal. The B52H has a length of about 49m, a wingspan of about 56m and is powered by eight Pratt&Whitney turbofan jets of about 75kN each. It carries a payload of 23,410kg and has a range of over 16,000km. Its maximum speed is 957km/h and it normally cruises at 820km/h. Modern bombers carry not only ordinary free falling bombs, but also cruise missiles and other high accuracy guided bombs. The most recent precision bombs are truly guided missiles that can hit their targets with fearful accuracy and fearful detonating power. Sometimes, unfortunately, they miss their target and hit something else with fearful detonating power. There is a miscellany of guidance systems, ranging from a computer-stored topographic map to laser beams to satellite navigation systems.

Another example of an American bomber is the supersonic Rockwell B-1B.This is a much smaller beast, with variable swept wings of only about 42m at minimum sweep and 24m at maximum sweep. The variable sweep is necessary to adapt the aircraft from subsonic to supersonic speed. It can fly at a speed up to Mach 1.25, which is 1,325km/h at high altitude. The B-1B is powered by four General Electric turbofans each rated at about 65kN, but the thrust can be increased by an afterburner to 137kN. The payload is about 13,000kg and the range with a typical weapon load about 5,500km.

Equally impressive are the figures for fighter aircraft. One of the best-known US fighters, used by many air forces all over the world, is the Lockheed Martin F-16. It is powered by a single General Electric turbofan engine producing about 129kN thrust (or a similar Pratt&Whitney engine) and has a speed of 2,120km/h at high altitude and 1470km/h at sea level. Its range is limited to about 550km, but it can be refuelled in flight from special refuelling aircraft. It is equipped with a cannon and can carry a load of various bombs and missiles. One should properly refer to these machines as fighter-bombers.

Another well-known fighter is the McDonnell Douglas F-15. This is powered by two 129kN jet engines. It has a speed of Mach 2.5 and a range of up to 4,400km. It too is armed with a gun and can carry a variety of bombs and missiles up to a maximum weight of 11,100kg.

The social consequences of flying are enormous. Large airlines now carry millions of passengers. British Airways employs 53,000 people, carries 33 million passengers per annum and links 170 cities in 80 countries with the UK. Lufthansa carries 44 million passengers per annum and links Germany with 227 destinations in 88 countries.[12] Flying has become the common mode of travel over medium or long distances. Millions of people now use a second tier of airlines, the no frills cheap operators, mostly for non-business flights. Business, government, and academic conferences have become a way of life. Everybody attends conferences at frequent intervals and airlines, conference centres and hotels make a good living. The fact that there now is so much

---

[12] These figures are taken from G. Endres et al., 1998, Modern Commercial Aircraft, Salamander Books, pp. 197 and 204, and may now be out of date.

interaction between firms, between governments and between Non-Governmental Organisations (NCOs) on a global scale adds to the flurry of conference travel. Technology has undoubtedly contributed to this cooperation by providing rapid global travel and instantaneous global communications.

Millions of people take holidays in distant destinations that in previous years they could not even have dreamed of and mostly did not know existed. Tourism has thus become the main industry in many countries, including countries of Southern Europe, but even in countries situated in the Pacific or the Caribbean, and it has become an important industry in almost all countries. Seaside holidays in the South are within reach of people with even modest incomes. City tourism throughout the year has become important and winter sports are now indulged in by very large numbers of people. The consequences of all this are massive hotels, seaside resorts looking like big modern cities, ski-lifts on an unprecedented scale and grave dangers to the environment in many popular destinations. Natural resources, especially water resources, are often stretched well beyond their limits. The social role of wealthy tourists in poor countries is a contentious and debatable issue.

As so many major technologies, the air transport industry forms a technological system. The aircraft is only one part of the system, albeit the very central one, and the whole system has undergone substantial technological and organisational development. The airline industry requires a vast infrastructure. It begins with the design and manufacture of aircraft and their parts and control systems. It includes the massive construction of airports and the complex operation of running airports with aircraft landing and taking off at incredibly short intervals. The airports need a ground transportation system for passengers, employees, baggage and massive supplies, including aircraft fuel. The high density of air traffic is made possible by an elaborate system of computerised air traffic control. Because aircraft have three degrees of spatial freedom – meaning that they move in three-dimensional space - they could not possibly avoid collisions without the rigid discipline enforced by air traffic control. And, unlike motorcars, even the slightest collision of flying machines leads to fatal consequences. If every aircraft flew wherever it wanted, total chaos would result. The problem of control is worse than for cars that have only two degrees of freedom, though that is quite severe enough. Because large numbers of aircraft could not possibly be controlled, the mass ownership of aircraft never became a serious proposition. In a sense, air traffic consists mostly of public transport with very few private planes thrown in for good measure. Thankfully, most of these fly at much lower altitudes from small airfields.

The motivations that caused aircraft development are illustrative of motivations for technological development in general. The early pioneers were fascinated by the dream of flying. Since the dawn of humankind humans must have felt envy for birds and their freedom from the shackles of gravity. We are tied down to earth, and we dream of being as free as the birds. The most famous of the many ancestors of this dream are the legendary Icarus and Leonardo da Vinci – one who flew but paid with his life for his presumption, the other who toiled and dreamed but never flew. The pioneers who actually flew must have been inspired by these very same dreams. As all technological pioneers they attempted to achieve what no one had achieved before, but in their case the desired achievement was a strongly felt human desire, shared by all humankind.

The second motivation (or was it the first?) was, as always, the desire for fame and fortune. The Wright brothers, for example, felt a strong urge for technological achievement, but they also were no mean businessmen and, after their initial success, went into the business of making money out of building aircraft. The same was true for all the other early, and later, entrepreneurs and their financial backers.

In the case of aircraft, the military became involved quite early on. It was obvious to any military person with imagination and foresight that here was a machine that could bestow military advantages to its users. Even the balloon was used for reconnaissance and the earliest powered aircraft were used first for reconnaissance and later for bombing or strafing enemy lines or strategic targets. For those who wish to argue that technology develops only in response to human needs, the fact that the military decided to use aircraft at an early stage provides an argument in favour of their belief. In my view, the military need came after the event. The initial motivations for the development were, to simplify the argument, curiosity, vanity, and greed. The apparent need was an afterthought; indeed it was the realisation that here was a technology that might have its practical uses. And this, I think, is the most common sequence of events. Necessity is not usually the mother of technological invention, even if the idea is tempting and sounds plausible. The question the entrepreneur asks is "what can I sell", not "what do people need". It is willingness to buy, not need, that determines the success of a product. In our rich Western societies there is little congruence between our purchases and our needs. The distinction between need and want is a theoretical one, in reality people buy goods to satisfy their wants and these include their essential needs.

Electricity

Although electricity is perhaps the most characteristic and most fundamental feature of 20th century technology and has the most far-reaching social effects, we shall introduce it in only the briefest of terms. Though lighting can be provided by other means, such as by town gas, electric power has no substitute and no equal. All manufacturing machinery is now driven by electric motors, all domestic appliances rely on electricity, all computers, radio and television rely on electricity, all water supplies and even domestic heating appliances need electric power. From razor to washing machine, from industrial robot and steel mills, from air conditioning to elevators, from whatever comes to mind to whatever comes to mind, they all need electric power. It is the ubiquitous power of our century; only mobile applications such as motorcars or aircraft rely on other forms of energy.

We start our narrative with electric lighting and with possibly the best-known inventor of all times. Thomas Alva Edison announced the carbon filament light bulb in late 1879 with a great deal of publicity. Edison had his own private industrial laboratory in Menlo Park, perhaps the first R&D laboratory dedicated specifically to technological innovation. The rate of patenting was prolific: in a single year, 1882, Edison applied for 141 patents! The work of Menlo Park included improved light bulbs, eventually substituting tungsten for carbon and thus obtaining much brighter light. It also included heavier electrical engineering, such as generators with greater efficiency, transmission of electric power and electricity meters. In the same year, the Edison Electric Illuminating Company built its first central power station in Manhattan. There was severe rivalry between gas and electricity for lighting and much of the rivalry was based on competing fears. People feared electricity because it was generated by steam engines and thus required steam boilers that might explode; gas, on the other hand, might leak and cause death by poisoning or by explosion. Eventually electricity prevailed because of greater convenience, a greater variety of applications and, last but not least, because steam boilers operating in remote power stations became perfectly safe.

Before electricity won completely, it had to go through a period of another intense rivalry, this time between direct (dc) and alternating current (ac)[13]. Edison used and advocated dc and his initial lighting installations and power stations produced dc. The original invention of dc generators and motors was made in the late 1860s by Zénobe-Théophile Gramme in France.

Till 1888 there was no alternative to dc for generating electricity and for driving electric motors. Though Lucien Gaulard and his English business partner John Gibbs started manufacturing transformers in 1883, making it possible to change ac voltage to higher or lower values with very little loss, these did not become truly useful till 1888, when Nikola Tesla, an American of Croatian origin, patented an ac generator and motor, thus removing the last obstacle to the general use of ac. The great strength of ac is the fact that its voltage can be changed so relatively easily. Hence, long distance transmission can be carried out at a high voltage. Because power W is the product of voltage V and current I, transmission at high voltage means that the current can be kept relatively low and this means that losses in transmission, proportional to the square of the current, can be kept low.

A classic case of technological and commercial rivalry unfolded when George Westinghouse bought Tesla's patent rights and his firm, the Westinghouse Electric Company, supported and manufactured ac devices. Edison strenuously opposed them. There were court battles over patent rights in the lighting field and commercial as well as scientific battles over superiority of ac or dc systems. It took many years for a decisive victory of ac to emerge. Edison himself lost much of his influence when the various Edison companies merged into the Edison General Electric Company in 1889, then merged with the Thomson-Houston Company in 1892 to become the great General Electric Company. Peace with Westinghouse was achieved in 1896 with a patent exchange agree-

[13] We speak of dc, or direct current, when both conductors retain their positive or negative polarities at all times. We speak of ac, or alternating current, when the conductors swap their polarities cyclically.

ment between the two companies. A uniform electrical system gradually emerged in the USA from the many incompatible electrical systems, both ac and dc. Before the USA settled on the 115 Volt standard, another spate of fear had to be overcome. The fear of exploding steam boilers was replaced by the fear of electrocution. It was believed that high ac voltage was particularly lethal and that the supply voltage should be kept well below a lethal level. The agreed 115 Volt for the US electricity supply emerged as a compromise between fear and economy.

In the United Kingdom the incandescent filament light bulb was invented independently by Joseph Swan in 1860, and in practically usable form around 1880, and manufactured by his company along with a variety of other electrical goods.

In 1902 there were 258 electricity-generating stations in the UK; of these 59% were municipal and 41% private. There were no standards. Some stations produced ac, some dc, and the voltage and frequency (in the case of ac) was different for each station. In 1917 there were 70 different generating authorities in London, using 50 different systems, 10 different frequencies and 20 different voltages. As electricity generation in the early stages was tied up with lighting, it was regarded as a local independent affair and nobody in the early days envisaged the enormous importance of electricity and the need for standardization.

In order to use the most powerful electric motors economically, it is necessary to produce a three-phase ac supply. This consists of one neutral conductor and three conductors each at a voltage that was later standardized to 240 Volt in relation to the neutral conductor and 380 Volt in relation to each other. The real point of the three-phase supply is that the voltage rotates and thus provides a rotating magnetic field that powers the induction motor. The first such supply was provided in London in 1900. For most large electric motors a three-phase supply is now used, the only exception being the dc motors often used for electric traction.

In 1919 Parliament passed the Electricity Supply Act that established Electricity Commissioners with the duty of "promoting, regulating and supervising the supply of electricity". The state had woken up to the importance of electricity and to the need for some form of state intervention. In 1926 a new Electricity Supply Act was passed and it was decided to establish a Central Electricity Board with the task of constructing a National Grid. Construction work started in 1927. By 1935 almost all of Britain was linked to 4600km of primary transmission lines and 1900km of secondary lines, linked to 642 electricity supply undertakings. Standardization to 3-phase 50Hz was not completed till 1947 and the voltage was standardized to 240V in 1945. The Central Electricity Generating Board was established when electricity supply came under public ownership in 1948. This lasted for about 40 years, when electricity supply and distribution came into private ownership again. The example of electricity, a typical technological system, shows clearly that state intervention is of the essence to prevent incompatibility of parts of the system and thereby greatly reducing its utility. State intervention is also vital to control the not inconsiderable hazards posed by electricity used incorrectly.

Telecommunications

Air travel has contributed a great deal to the shrinking of the world. Technology has, in a very real sense, eliminated distance. But there is another factor contributing to the elimination of distance: telecommunication. As the name implies, this is communication at a distance, instant and without restrictions owing to distance, from any point on earth to any point on earth or beyond it. Distant communication is an essential need of the military and has been tackled in various ways since time immemorial. Communication, whether at a distance or by messenger or letter or any other means is vital for many civil activities and indeed vital for the very existence of society. We need to communicate in trade of every kind, we need to communicate to promulgate laws or collect taxes or carry out any social cooperative activity. As society becomes more extensive and more complex, so the need for communication increases and speeding up the rate at which information can be transmitted offers real advantages, even though it must be said that the present emphasis on instant transmission of tremendous masses of information is largely vacuous and motivated by greed rather than by need. Technology is only interested in the mode of transmission of information, not in the content of what is being transmitted.

Hand signals must have been part of cooperative hunting and of cooperative fighting from the earliest days of humankind. Somewhat later, bonfires on hilltops or smoke signals were used to convey simple messages over great distances. Ships used hand signals or flags; armies used drums and bugles as well as hand signals. Couriers have been used since the earliest civilizations. In the 19th century, organised lines of semaphores became standard signalling practice. In France there were 3,000 miles of such lines in 1840, all operated by the War Department. In Russia, a semaphore line, consisting of towers five or six miles apart, connected St. Petersburg to Warsaw and beyond.

Early in the 19th century electricity began to be understood and it was soon realised that electrical signals could be used to transmit messages over wires. Two eminent German scientists, Wilhelm Eduard Weber and Carl Friedrich Gauss, may be regarded as inventors (albeit probably not sole inventors) of the electrical telegraph. In 1833 they used a magnet and a coil of wire and, by moving the coil in the magnetic field or by changing the magnetic field, induced an electric current in the coil (an effect discovered by Michael Faraday in 1831). This current they transmitted to the other end of town through a pair of copper wires slung over the church steeple at Göttingen. The current was detected at the other end by the movement of a magnetic needle within a coil of wire, an effect discovered in 1820 by Hans Christian Oersted. The discovery that electromagnetic signals could be produced, transmitted, and detected was of fundamental importance but did not lead directly to a practical device. The experiment by Gauss and Weber was part of extensive investigations into the properties of electricity and magnetism, pursued at that time by many scientists in many countries. This first experimental era culminated in the formulation by James Clerk Maxwell of the laws of electromagnetic fields in 1864.

The electromagnetic system of telegraphy became practical only with an agreed code for the interpretation of the signals into language. Many inventors worked on such a code and on devices for recording the messages received and many disputes over patent rights ensued. Because the purpose of patents is to grant the inventor a temporary monopoly on the use of his invention, and thus to safeguard the financial interests of the inventor, patents are often subject to litigation. In my interpretation this is further proof, if such proof is needed, that technological inventions are brought to market in order to make money. The desire for profit is the number one motivation for technological invention and innovation, even if other desires also play a role. The most successful inventor of practical telegraphy was Samuel Morse, who perfected the telegraph in the years 1832 to 1835 and finalised the Morse code – a system of dots and dashes standing for letters and numerals – in 1838. The code can, of course, be used for signalling with lights or flags or whatever, but the really significant method is electromagnetic telegraphy in which electrical signals are transmitted and recorded. The transmission was originally by wire, but with the invention of wireless transmission, wireless telegraphy became an alternative. The original Morse code proved inadequate for some languages other than English, and a European conference, convened in 1851, devised a modified code that became known as the International Morse code and is in use to the present day. The first public telegraphic message was sent in 1844 and telegraphy went from strength to strength for more than a century.

The first users of telegraphy were the railways and the newspapers. Railways needed telegraphy for operational reasons; newspapers desired it in order to speed up the spread of news. As the importance of telegraphy for the military was seen from its beginning, a debate about ownership of telegraph rights began. The solution to the ownership question varied from country to country, depending on whether more emphasis was put on private profit or on public interest.

The next step in the development of telecommunications was to add the transmission of speech to the transmission of written messages. The nature of speech as a pattern of waves carried by air became understood in about the middle of the 19th century. The science of acoustics was born and brought to a first climax by Hermann von Helmholtz, a physiologist and physicist, in his classic book of 1863. Helmholtz, a firm believer in the empirical foundation of science and opponent of purely philosophical scientific theories, began his career as a physician in the Prussian army and ended it as the first director of the Physico-Technical Institute[14] in Berlin, founded in 1888. He was thus a key figure in the emerging symbiosis between science and engineering.

Some years earlier, in 1855, Édouard-Léon Scott de Martinville had produced the phonautograph (soundwriter) consisting of a large cone with a membrane. A pig's bristle was attached to the membrane and wrote on a blackened glass that was moved along. Thus a trace of speech spoken into the cone was obtained. Though of

[14] This was a standards and research institution run by the state

no immediate practical consequence, this device must be regarded as a predecessor to the phonograph invented by Thomas Alva Edison in 1877. We may look upon the problem of transmitting speech as closely associated with the process of recording speech. What need to be done is to transform speech, i.e. acoustic vibrations, into an electrical signal, transmit this signal, and reconvert it into an acoustic signal.

Alexander Graham Bell, originally from Edinburgh, settled in Boston in 1871 as a teacher of speech and elocution, with particular emphasis on teaching deaf children to speak. He was interested in acoustic apparatus and shifted his attention to the intriguing possibility of telephony. In 1876 he patented a device for the transmission of an undulatory electric current, now generally regarded as the basic invention of the telephone. An almost simultaneous patent application for a similar device by Elisha Gray led to a prolonged and bitter battle over priority of the invention and rights to exploit it. The circuit described in Bell's first patent is somewhat weird and wonderful. Apparently the electrical signal was to be modulated by immersion of a metal wire in mercury, thus altering the resistance of the circuit. In a second patent, of 1877, the circuit consists of a small magnet attached to a membrane and moving in a wire coil, thus inducing a current in the coil that follows the movement of the membrane. The receiver is simply the reverse of the transmitter: the modulated current in the coil causes the magnet and membrane to move. Attaching the coil to the membrane and leaving the magnet stationary can achieve the same effects. In Bell's arrangement the transmitter and the receiver were identical. The principle of the moving magnet is one possible design for a transmitter, though modern telephone transmitters (microphones) are based on a patent filed by Edison in 1878, using the fact that the resistance of carbon powder can be modified by pressure. If a capsule is filled with carbon granules and is closed by a membrane, the moving membrane modifies the resistance, and thus the current in a suitable circuit. This type of microphone went into production in 1895 and has been widely used. In this case the receiver, based on the principle of the moving coil or magnet, is different from the transmitter. In modern parlance we would say that the telephone is an analogue device because the current induced in the microphone closely follows the movement of the membrane. In other words, the current is analogous to the pattern of mechanical movement caused by the air-pressure waves of speech.

As far as the private citizen is concerned, the telephone falls into the category of want that might be called the "would it not be nice if" want. Nobody needed the telephone, but the thought of being able to speak to friends and family at a distance is a nice thought. For business users, the utility of the telephone was far more obvious. There were real benefits to be had from the ability to speak to managers located in different locations of the same company, or to suppliers, or to customers. At a time when companies grew and began to spread their operations to different locations, the telephone was useful to such an extent that it might be classed as a real need or, more accurately, as a factor that determined the way businesses operated. Whether it increased their efficiency or not is a moot point. Undoubtedly they would have found modes of operation without the telephone, but the telephone did shape the corporations to a certain extent and for corporations operating in this particular mode, the telephone became indispensable. Major technologies used in organisations have a great influence on their modes of operation. Bell foresaw the possibility that business organisations might see benefits in the telephone and he and his associates and financial backers aimed the telephone specifically at the business market. The private market came much later, as a kind of afterthought.

Though eventually Bell came through with flying colours, he was involved in lawsuits for many years. The most difficult battles were fought against Elisha Gray who filed a patent for the telephone only hours later than Bell. The dispute was finally settled when the National Bell Company (Bell's company) appointed Western Electric (Gray's company) as its equipment manufacturer. Indeed Western Electric remained the manufacturing arm of Bell till long after the latter had become AT&T.

The first small manual telephone exchange, serving five banks, was installed in Boston in the spring of 1877. The total number of telephone users in New York reached 778 in the same year, including several stockbrokers. In 1889 Almon Strowger invented a mechanical switching device that was first introduced in 1892. The automatic telephone exchange, at first using the Strowger switch, spread widely when the rotary dial was added to telephones from 1896 and, in standard form, from 1910. From these small beginnings both telephony and telegraphy went from strength to strength. It was telegraphy that managed to span the oceans first by laying submarine telegraph cables.

The Transatlantic Cable[15]

News travelled slowly across the oceans, as it was limited by the speed of ships. As telegraphy spread within Europe and America and other continents, the idea of laying cables across narrow stretches of water within the continents naturally presented itself. Indeed the first cable connecting England and France was laid in 1851 and in the following few years England became connected to Holland and Ireland, and Italy became connected to Corsica and Sardinia. Telegraphy had become important to newspapers as well as to civil government and the military. It helped to control far-flung government, business, and military operations.

In 1854 the wealthy American Cyrus Field decided to organise the laying of a telegraph cable across the Atlantic Ocean, essentially connecting Europe with the USA. Field's wholesale paper business could manage perfectly well without him and did not provide the challenges that Field desired. He was a man with a penchant for heroic deeds and there was nothing heroic about selling paper and printing supplies. He was also a man fond of unusual business enterprises.

Field contacted a number of people that he thought might be helpful, mainly wealthy potential investors in the enterprise, but also experts in oceanography and telegraphy. Morse, one of the latter group, promised his support and the oceanographers assured Field that conditions on the route from Ireland to Newfoundland were favourable, with the ocean bed reasonably smooth and flat, and the depth between 1500 to 2000 fathoms (approximately 2700 to 3600 metres). The investors also proved interested and Field founded the New York, Newfoundland and London Telegraph Company and raised a capital of $1.5 million. This was obviously a business enterprise. The aim of the investors was to make a profit, and this meant that they were keenly interested in the success of the enterprise. None of them were technically educated, but all of them appreciated the business benefits this technological enterprise might bestow upon business in general and their own business in particular. To most of the investors the transatlantic cable was an investment opportunity like any other; for Cyrus Field it became a passion.

The first step was to construct a line between New York and Newfoundland. The Governor and Assembly of Newfoundland granted the company a charter and gave it considerable financial support in the hope of bringing business and employment to Newfoundland. The line was duly constructed, but it took a whole year and devoured a third of the company's total capital. Part of the line had to be laid over rugged terrain and one stretch of 85miles had to cross the sea. After an initial failure, this stretch of submarine cable was manufactured and laid successfully by a British company, using the steamer Propontis.

Cyrus Field was the untiring driving force behind the enterprise and made numerous trips to England to raise political, financial and technical support for the project. In 1856 he founded a second company in Britain, the Atlantic Telegraph Company. The company obtained its charter and a promise of generous help from the British Government. With a huge empire to govern, it was very much in the British interest to obtain rapid and reliable communications with its far-flung territories. Good communications would be helpful to the administration, to business and, last but not least, to the military. Although the USA was not yet a world power with global interests, extensive lobbying by Cyrus Field and his friends extracted a matching promise of support from the American Government.

Both British and American naval vessels surveyed the proposed route for the cable and conditions were again found to be favourable to the enterprise. The actual design of the cable proved to be rather controversial. Every known authority on matters of electricity, especially matters of the propagation of signals in an insulated cable operating under water, was consulted, but the advice was not unanimous. The eminent physicist William Thomson, the later Lord Kelvin, and others advocated as thick a copper wire as practicable in order to reduce resistance to a minimum. The equally eminent Michael Faraday advocated a thin cable in order to reduce capacitance. The fact of the matter was that the problem was not fully understood and that even concepts such as resistance and capacitance were somewhat vague. The board of directors of the company, faced with contradictory scientific advice, did the obvious thing and chose the thin cable because it was a lot cheaper.

---

[15] This description is based almost entirely on the fascinating book by John Steele Gordon, 2002, A Thread Across the Ocean, Simon & Schuster.

The cable was duly manufactured. It consisted of seven strands of thin copper twisted together. The insulation consisted of three layers of gutta-percha[16] followed by a layer of hemp saturated with tar, oil and wax. Finally, the cable was armoured with iron wires twisted round it and coated with another layer of tar to protect the wire from corrosion. Thomson tested the cable and found that the copper was not nearly as pure and as homogeneous as he had wished. The total length of cable of 2,500 nautical miles (4,600km) weighed 1 ton per mile. This made it too heavy to load on any existing ship and thus had to be loaded in two halves on two ships. The two ships chosen were the USS Niagara and the HMS Agamemnon. It took 30 men three weeks to load the cables and the two ships set out with a naval escort in the summer of 1857. The cable broke when about 400 miles had been payed out because the paying out machinery was inadequate to the task and a mechanic had failed to release its brake when the ship rose on a wave. Thus 400 miles of cable were lost and the attempt was abandoned, to be restarted in the following year.

A new chief engineer was appointed, a William Everett, on leave from the US Navy, and new paying out machinery was built to his design. The design has stood the test of time and has survived, virtually unchanged, to the present day. The cable was frequently tested by sending a Morse signal from one end and picking it up with a sensitive galvanometer, specially designed by William Thomson, at the other end. The second attempt at laying the cable was started in mid-Atlantic, with the two ships sailing in opposite directions. The attempt was fraught with difficulties. The ships encountered a most terrific storm and the cable broke a few times, but could be spliced again with only short lengths lost. Eventually the cable snapped and a new attempt (the third) was restarted in July of the same year, 1858. This time the cable was laid successfully from end to end. To celebrate the occasion, Queen Victoria and President Buchanan exchanged messages on 16.August. The truth was, however, that the cable worked very poorly. Signals were weak and transmission slow and after a few weeks the cable went dead altogether.

Cyrus Field was a man of great perseverance, bordering on sheer obstinacy. He managed to get a Commission of Enquiry appointed that was to look into the causes of the failure of the three attempts to lay the transatlantic cable. Leading scientists, including Sir Charles Wheatstone and Sir William Cooke, joint holders of a telegraph patent of 1843, participated in the enquiry and made numerous recommendations. One of the results of their recommendations was the proper definition of concepts and units, such as the watt, volt, ohm, ampere and more. They also recommended that the cable be properly tested before and during the laying operation.

Another unrelated and independent technological development proved crucial to the eventual success of the transatlantic cable. Isambard Kingdom Brunel had completed his – by the standards of the time – gigantic and revolutionary steamship - the Great Eastern. It was launched in January 1858 (after a previous failed attempt) and had had a rather lack-lustre career. She was too big and too advanced for her time and several owners had tried in vain to make money by using her as a passenger and cargo ship. Brunel died during her sea trials, but not before having drawn Field's attention to the possibility of using her as a cable-laying ship.

Field went about the difficult business of raising more money for another attempt at laying the elusive transatlantic cable. By 1864 he had secured sufficient finance and sufficient political support to be ready for another attempt. A new cable, much heavier than the previous one, was designed by the physicist William Thomson and the engineer Charles Bright. It was manufactured by a new company, formed by mergers, the Telegraph Construction and Maintenance Company. Many improvements had been made. Apart from the wire having a greater cross-section, the copper it was made of was purer. Both these measures reduced the resistance of the cable. The insulation consisted of four layers of gutta-percha supplemented by a new insulating substance, Chatterton's compound, between the layers and over the wire. Then came a layer of hemp soaked in pitch and finally an external armour made of good quality steel wire, coated with hemp. The cable was tested at every stage and completed at the end of May, 1865. Field was now the only American involved in the enterprise, all other participants, and most of the capital, were British. Only one journalist, William Howard Russell of the

---

16 Gutta-percha is obtained from a tree native to Malaya and is similar to rubber, but not as elastic. It is thermoplastic and sets to a hard but somewhat flexible insulating substance. Some years earlier Werner von Siemens had constructed a machine for coating wire with gutta-percha, thus producing cables.

Times, was on board when the Great Eastern set sail on 23rd July from Ireland on its transatlantic voyage. This was the fourth attempt.

It was not all smooth sailing. Some bits of wire were found stuck in the insulation, shorting the cable. At one stage the cable had to be retrieved from the ocean bed to repair a fault. Finally, the cable snapped, disappeared into the ocean and was not found again till a year later. Most men would have given up, but not Cyrus Field.

A new company, the Anglo-American Telegraph Company was formed and a new cable was manufactured. This time the outer armour was made of zinc-coated steel, thus preventing corrosion. The outermost layer of hemp was not soaked in pitch and was not sticky, as its predecessors had been. The new expedition left Ireland for Newfoundland on 13. July, 1866. This time all went smoothly and the end of the cable arrived in Newfoundland a fortnight later, on 27. July, 1866. The cable that had been lost previously was raised from the ocean and laid successfully, as a second cable, in September, 1866.

Cyrus Field died in 1892, having been involved in several other exciting business projects. The traffic density on the cable was initially low, because prices were exorbitant. One of the results of this initial pricing policy was the formation of United Press International, which allowed newspapers to share the cost of transatlantic dispatches. Eventually prices dropped and traffic increased. The operation was profitable throughout and the laying of submarine cables became fashionable. By 1900 there was a total of 15 transatlantic cables in operation, as well as a cable from Suez to Bombay. Australia was reached in 1871 via Singapore, and China and Japan followed. In 1902 a line from Vancouver to Australia and New Zealand completed the cable-girdle round the globe.

In 1914 the British Navy successfully cut submarine cables connecting Germany to the world and thus forced the Germans to use wireless communications during World War I, thus becoming more susceptible to having messages intercepted.

The original submarine cables proved suitable for telegraph communication, but not for telephony. The rate of transmission was too slow for that. Submarine cables with so-called repeaters, or regenerators, i.e. amplifiers that regenerated the signal at intervals along the cable, were first introduced in 1956. Cables were then capable of carrying 33 simultaneous telephone calls. This figure rose to many thousands, and even hundreds of thousands over the following decades, especially with the introduction of fibre-optic cables. Despite current competition from wireless and satellite connections, much telephony is still carried by cable, though the major signal carrier is now an optical fibre carrying light pulses rather than a copper wire carrying electrical pulses. The numbers of international calls made annually from the USA has risen at a staggering rate: In 1950 it was one million; in 1997 it was 4.2 billion calls.

The meteoric rise of telephony began when Bell started introducing automatic exchanges (central offices) on a major scale immediately after the First World War, in 1919. But it was the universal introduction of direct dialling, with every subscriber in the world able to dial every other subscriber without human intervention, that spread rapidly in the post World War II period, and particularly in the 1970s, that made the telephone an instrument of global communications. With rising prosperity in the same period the telephone became ubiquitous and, in the Western World at least, ownership of telephones is now almost universal.

We are ahead of ourselves and have to return to the early days of telegraphy and telephony. It followed from Maxwell's theory that electromagnetic phenomena should spread through space much like waves spread in water. In other words, an oscillating electric charge with its associated magnetic field, i.e. an electromagnetic field, spreads from its point of origin as a wave. The predicted phenomenon was experimentally confirmed by Heinrich Hertz in a series of experiments conducted in the years 1885 to 1889. Hertz also demonstrated that light consisted of electromagnetic waves. The difference between radio waves and light or heat waves is merely in their respective wavelengths.

The Russian Aleksandr Popov was one of the first to transmit and receive electromagnetic radiation, radio waves, over some distance and built a primitive radio transmitter and receiver in 1895. In Russia he is considered to be the inventor of radio communications.

In the West, the same role is assigned to Guglielmo Marconi, who experimented with radio waves and filed a patent for radio telegraphy in 1896. The two inventors did not know about each other. The transmitter consisted of a high voltage spark that was controlled by a Morse key. The receiver consisted of a weird device known as a coherer, perfected earlier by Sir Oliver Lodge. The coherer consisted of a glass tube with two electrodes, filled with loose iron filings. Normally the tube does not conduct electricity because the filings are packed too loosely. In the presence of an electromagnetic wave the filings cohere (as shown in 1890 by Édouard Branly) and render the tube conducting. If the tube is part of an electric circuit, a current will be recorded. If the tube is immediately struck by what was called a trembler, the filings become loose again. In this way the coherer can be used to detect electromagnetic waves and was indeed used by Marconi for this purpose. Lodge and Muirhead had transmitted and received an electromagnetic signal over a short distance in 1889, merely for the purpose of demonstrating electromagnetic waves. Marconi was not interested in demonstrating a scientific fact – he was after the practical possibility of signalling over a distance without using a wire. For obvious reasons, both Marconi and Popov aroused the interest of their respective navies. Marconi improved the performance of his apparatus greatly by adding an aerial and in December 1901 managed to transmit a signal from Newfoundland to Cornwall. The spark gap as transmitter and the coherer as receiver were crude devices, just sufficient for the transmission of Morse code. Great improvement in the range of broadcasting was achieved by implementing an invention by the German physicist Ferdinand Braun of a novel antenna circuit linked to the transmitter circuit by induction, patented in 1899. The 1909 Nobel Prize in physics was awarded jointly to Marconi and Braun for their invention of radiotelegraphy.

Though Marconi achieved some commercial success with these devices, he filed another patent in 1900 in which he introduced the possibility of transmission at predetermined wavelengths. Marconi's second patent was based on previous work by Oliver Lodge, Nikola Tesla and others and was eventually overturned by the US Supreme Court. However, the principle was established that a tuned circuit – tuned with the aid of variable inductance and capacitance –could be used to determine the wavelength of a broadcast radio signal. The tuned circuit has remained a crucial element of radio communication.

For the transmission of complex information by radio waves, we now use a so-called carrier wave that is modulated – either in amplitude or in frequency – by the signal to be transmitted. The receiver is tuned to the wavelength of the carrier wave and detects the signal carried on the wave. The detector – essentially a rectifier – was invented and improved during the half century following Marconi's initial success. Different types of rectifier, based on different physical phenomena, have been used.

The first type of rectifier is based on the vacuum diode invented in 1904 by Sir John Ambrose Fleming. The principle is simple. If we take a light bulb, consisting of a hot filament in an evacuated glass envelope, add a second unheated electrode, and apply a voltage between the heated electrode (the cathode) and the cold electrode (the anode) current will flow only if the anode is positive and the cathode negative. If we apply an alternating current, only that part of the current will flow through for which these conditions are fulfilled, thus the alternating current will be converted to direct current, we say that the valve rectifies the ac current. The reason is that the heated cathode emits electrons, carrying a negative charge that can be attracted to the anode, whereas the anode does not emit electrons. If a high frequency current is received in a circuit containing a rectifier, only the dc component will flow through and this can be measured by an instrument such as a galvanometer, or it can drive a suitable receiver.

In 1906 Lee de Forest inserted a grid, a second electrode that allowed the passage of electrons through it, between cathode and anode. A voltage applied to the grid could greatly influence the current flow through this triode valve, (initially called Audion by its inventor) and the valve could thus act as both rectifier and amplifier. In the following years vacuum valves (vacuum tubes in American parlance) with further grids were invented, the tetrode, the penthode, and so forth. Whereas Fleming was an academic scientist, de Forest was a scientist/businessman and was determined to be seen as the father of radio. A fierce battle over the priority of invention between them is an unfortunate characteristic of invention and innovation. When money and/or fame can be made from priority of inventions, protagonists are bound to fight over priorities, especially as inventions of a closely similar character often occur at roughly the same time. Once the preconditions of knowledge and technology are established, fertile minds attempt to take the next logical step of development and thus bunching and simultaneity of inventions can easily occur.

The various inventions of valves, of fixed and variable capacitances, inductances, resistors and transformers, not to mention aerials, made the broadcasting radio a practical proposition. A whole new branch of engineering, electronic engineering, developed during the next half century, based entirely on these types of active and passive circuit elements. Radio was one practical result, a variety of amplifiers and servomechanisms became possible and, eventually, early computers and television resulted. In the second half of the 20th century electronic engineering turned mostly away from thermionic valves and replaced them by solid-state electronics. The transistor, invented in 1947, and the silicon integrated circuit led to an enormous diversification and expansion of electronic engineering.

In the early days of radio, with Lee de Forest one of the leading lights, radio broadcasts became established. Perhaps the most important contributor to this development was David Sarnoff who worked his way up from messenger boy to radiotelegraph operator and achieved his first bit of fame when he picked up the distress signals from the sinking *Titanic*. He was soon promoted within the Marconi company and began to promote the idea of broadcasting entertainment on radio and marketing a radio receiver. In 1921 he was manager of the newly formed RCA (Radio Corporation of America). The broadcasting of the commentary on a heavyweight-boxing match proved tremendously popular and RCA sold radio receivers in very large numbers. In 1926 Sarnoff founded the NBC (National Broadcasting Company) and thus completed the establishment of radio as a popular means of receiving news, other information and entertainment. Radio has never looked back. By stressing the role of one man, however important, we do an injustice to the many other people involved in as large an enterprise as the establishment of broadcasting. But even a detailed history cannot do justice to every contributor, and this is only a short sketch, far short of a detailed history.

The same apology has to be made when we come to the subject of television. A very large number of inventors described a variety of methods that might be used to broadcast moving pictures. The Russian scientist Boris Rosing described and built a prototype of a television receiver using a cathode ray tube, invented in 1897 by Ferdinand Braun. In 1907 he managed to send and receive crude pictures and may thus be regarded as one of the successful pioneers of television. Vladimir Zworykin, a Russian-born American engineer, is often regarded as the main inventor of modern television, though he is by no means the only contender. Zworykin became a researcher in the Westinghouse Electric Corporation in 1920. In 1923 he patented a television transmitter tube and in the following year a receiver tube. In 1929 he demonstrated an improved system and moved to RCA as head of electronic research. In the following years Zworykin and his team made further improvements in both black&white and colour television. Research into television was active outside the US as well. In Germany Baron Manfred von Ardenne demonstrated a television system in 1931. In Britain, John Logie Baird demonstrated a partly mechanical television system as early as 1925 and a research group at EMI, led by Isaac Shoenberg, produced its own fully electronic system in 1931. It had become clear that the definition of mechanically scanned systems was necessarily inferior to that of fully electronic systems. Modern television was the child of many parents; virtually all of them members of research groups in industrial laboratories. The period of systematic industrial R&D had come into full swing.

World War II delayed the spread of television, but it regained momentum rapidly in the post-war period. Television-broadcasting organisations were set up in most countries and the number of programme channels increased. When the market for black&white receivers became saturated, colour television was introduced and thus flagging sales were revived with a better and more expensive product. Commercial broadcasting was introduced in many countries – following the American lead – and competition between the many channels led to ever lower quality standards. Zworykin himself, the leading progenitor of television, in later life lamented the way television had been used to trivialize subjects instead of for the educational and cultural enrichment of audiences. There can be little doubt that in this particular field competition leads to declining standards by tending toward the lowest common cultural denominator. In the search for mass audiences, anything that titillates or entertains at the least demanding level is welcome. Violence has become part of the daily fare of TV. There can be little doubt – even if some pundits do doubt it – that the routine constant showing of violence on television leads to imitation of violent behaviour among the uneducated young. To them, violence appears to be a normal pattern of behaviour rather than the uncivilized aberration that it is. Indeed the role models that television presents are often pathetic and lamentable rather than worthy of imitation. Some programmes on some channels are, of course, excellent. Some news, some discussions, some drama, some nature films and some science programmes are true enrichments for their viewers. But most television is harmless trash at best and dangerous rubbish at worst.

The spread of television has extremely far-reaching social consequences. The majority of people throughout the world spend a large proportion of their leisure time watching television. Sitting in front of the goggle box has largely replaced other forms of social life, including conversation within the family and casual visits by friends. Television is a window on the world and, because of its poor quality, it provides a very limited and distorted view of this world, especially as it is infiltrated by insidious propaganda. It has also led to the growth of the personality cult, a wholly deplorable cult in view of the fact that the cult figures are not usually worthy of admiration or imitation. The stars of today's television world glitter, but their glitter is not that of diamonds but of glass beads. Television has often been hailed as an instrument of democracy. It was thought that an informed public would be able to make informed choices in their voting behaviour and in their political activities. Instead, the public is as ill informed as ever and politicians are reduced to sound bites that are meaningless but memorable. Unfortunately, sound bites are no substitute for information or argument, and photogenic politicians are no substitute for thoughtful, wise and honest leaders. Most discussions are too brief and too lacking in discipline to be meaningful and often deteriorate into slanging matches. Despite all slogans on political independence, political patronage has proved too powerful and inconvenient views are rarely aired. Television influences public taste to a considerable extent. Unfortunately, it mostly goes for the garish and vulgar rather than for good quality stylish design.

Television has contributed to enormous growth in the entertainment and advertising industries. It has grown far beyond its predecessor and contemporary industries, such as film, radio and the press. Growth in the entertainment industry was an inevitable consequence of reduced working hours and increased leisure time. In itself, this is not a bad thing provided the entertainment is of good quality rather than vacuous pulp. The growth in advertising is a much more dubious proposition because it educates us to become consumers, it projects to us a make-believe world in which truth is irrelevant, and it attempts to make us dissatisfied with our lot, whatever that lot might be. To be happy, we have to buy, whether we need it or not. I think that advertising has induced a very cavalier attitude to truth and value in many of us.

The television industry has again reached a stage when the market for television receivers is saturated. Everybody has a colour television set. A new device is urgently needed that will revive the fortunes of the industry and will induce people to part with their money. It was thought that High Definition TV would provide the answer, but now the industry seems to have settled on digital TV instead and we shall soon all be buying digital sets or, at least, digital boxes to convert our existing sets to digital reception. The ultimate cry will be digital high definition TV using, of course, large flat screens rather than bulky cathode ray tubes. In return, we are promised better technical quality and more programmes. Will that make us happy?

Computing

The branch of electronics that has most influenced modern life, apart from radio and television, is computing. The predecessors to electronic computers are mechanical calculating machines of all kinds. Perhaps the most famous, though almost certainly not the most influential, of these machines is the analytical engine conceived by the English mathematician Charles Babbage. Babbage proposed two calculating machines: the "difference engine" and, in about1834, the "universal analytic engine". The latter was to perform any arithmetical operation, was to receive instructions on punched cards, was to be able to store numbers, and had sequential control. Thus the engine was to contain many elements of modern computers. Neither engine was ever completed in Babbage's lifetime. The Swede Georg Scheutz completed the difference engine in 1844, with an improved version in 1855. The machine was used for a few years to calculate actuarial tables for the British Registrar General. In 1991 some British scientists built Difference Engine No 2 on the basis of Babbage's notebooks. Babbage was a professor of mathematics at Cambridge University and pursued his ideas, with some government aid, for academic reasons; though it is likely that he, as most academics, always kept an eye on possible practical applications for the satisfaction of their pride and for possible financial gain. Babbage's engines, however, must be viewed as curiosities of no practical consequence.

There are many more predecessor mechanical calculating devices, some associated with illustrious names, such as Leibniz and Pascal. Of much greater practical importance were the various mechanical calculating machines that began to appear on the market in the closing years of the 19th century. The first key-operated calculator was demonstrated in 1887 and in 1893 William Burroughs added a paper roll and a printing mechanism to such a calculator. He founded the Burroughs Adding Machine Corporation in 1905 that eventually, in 1986, became part of the Unisys Corporation. For the analysis of the 1890 census in the USA the statistician Herman Hollerith designed a calculating machine that used inputs from punched cards, similar to those used by Joseph-Marie Jacquard in 1804 for the control of a weaving loom. Hollerith became the founder of the Tabulating Machine Company in 1896, which later became the International Business Machine Corporation (IBM). Mechanical or electro-mechanical calculating machines of all shapes, sizes and capabilities became firmly established in the administrative offices of commercial firms and in public administration and remained essential tools until they were displaced by digital electronic calculators and computers.[17]

A variety of miscellaneous electronic or electro-mechanical computing devices were built in several industrial research laboratories that might all be viewed as predecessors to the modern digital computer. These efforts began in the late thirties and continued into the forties. Serious large-scale computer projects also started in the 1940s in both USA and England. Two of the earliest electronic digital computers were built by John Atanasoff, later joined by Clifford Berry, at Iowa State College between 1937 and 1942. At Bell Laboratories George Stibitz used relays to build simple digital computers between 1937 and 1940. In Germany, Konrad Zuse built a series of computers in the years 1936 to 1949 that used electromechanical relays. The relays fulfilled the same function as the valves or transistors of later devices: they acted as binary switches with one position denoting a 0 and the other position the 1.

Atanasoff's ideas proved greatly influential and important for the most ambitious and best known of these early efforts, the ENIAC (Electronic Numerical Integrator and Computer), started in the University of Pennsylvania in the spring of 1943 and officially launched in early 1946. The project was supported by the US army in the hope that the machine would assist in new calculations of ballistic tables. The leading scientists involved in the project were John Mauchly, Herman Goldstine and J. Prosper Eckert Jr. The machine used more than 17,000 thermionic valves and was of enormous size. It is difficult to assess how urgent the task of calculating ballistic tables was, but there can be no doubt that some sort of military demand for the machine existed.

A more obvious and urgent demand existed in Britain for the specialist electronic calculating machine, the Colossus. The machine was specifically designed to help with the de-coding of German teleprinter messages in a system code-named Fish. Colossus, though designed for a specific task, had many features of a universal computer. It became operational in February, 1944 and made no small contribution to the cracking of German codes and thus to the British war effort. Colossus was just about the right name for it, for it used 1,500 vacuum tubes and must have consumed huge amounts of electricity. An enhanced version, Colossus II, used 2,400 vacuum tubes and became operational in June 1944. Colossus was built in Bletchley Park, an almost legendary wartime intelligence research centre. The popular fame of Bletchley, however, rests upon its successful de-coding of German messages using the Enigma coding device.[18] The best-known figures associated with Bletchley were Alan Turing, T. H. Flowers and Max Newman.

A milestone in the consolidation of computer design was The First Draft of a Report on an Electronic Discrete Variable Automatic Computer (EDVAC), written by the famous mathematician John von Neumann in June 1945. The report was partly based on the ideas developed by the team at the University of Pennsylvania in connection with ENIAC and a machine that was to follow it. This report, although in name a first draft, was widely circulated and discussed. In effect it laid the foundations to the design of modern digital computers. A further report on stored programme computers by von Neumann, in association with Herman Goldstine and Arthur Burks, further strengthened the foundations of computer design.

A flurry of activity followed the first computer efforts, initially driven almost entirely by cold-war military needs. The design of the hydrogen bomb required great computational effort and the growing activity of designing and building a variety of guided missiles involved no mean computational effort. In 1948 Nicholas Metropolis built a computer at Los Alamos Laboratory, following the von Neumann design. Jay Forrester built

---

[17] Much information on the history of the computer was gleaned from Brian Winston, 1998, Media Technology and Society, Routledge.

[18] I am much indebted to Peter Wolstenholme for putting me right on Colossus, referring to the Official History of British Intelligence in the Second World War (F. H. Hinsley et al.)

a similar machine at MIT. Princeton University completed a computer in 1952 that was used in the final design stages of the hydrogen bomb.

IBM began in earnest to build computers. One went to Northrop for guided missile work and almost all the others were used for military work. The only exception was one computer delivered to the US Weather Bureau.

In the late forties and early fifties it was still thought that a very small number of computers would satisfy all the foreseeable needs, mostly of a military nature. Only in the later fifties did it dawn on people that the computer could be more than a tool for the design of military hardware or ballistic tables or weather forecasts, and that it could perform useful tasks in commercial or state administration and in civil research.

Universities, in collaboration with the military or with commercial firms supplying the military, remained involved in computer design in the immediate post-war period. In Britain, a research group including Thomas Kilburn and Frederic Williams was active at Manchester University, designing the Ferranti Mark I computer. At Cambridge University, the EDSAC (Electronic Delay Storage Automatic Calculator) was built in 1949 under the leadership of Maurice Wilkes. The National Physical Laboratory (NPL) was designated to cater for computer needs at the national level in Britain. The NPL recruited the mathematician Alan Turing, of Bletchley fame, to help with computer design. Turing started building a machine, but it apparently was not completed and he left for Manchester and died soon after the move. NPL built a smaller version of Turing's design in 1950, called Pilot Ace. English Electric built the machine commercially under the name Deuce.

In USA, the National Bureau of Standards also became involved in computers and built the SEAC in 1950, which remained in operation till 1964. This machine used the newly developed solid-state diodes combined with thermionic amplifier tubes. This was the beginning of a synergetic relationship between the nascent semiconductor and computer industries. It did not come to full fruition, however, till the 1970s.

The rise of the semiconductor industry will not be described here; the interested reader is referred to an earlier book[19]. The development of the computer industry will be described, albeit with extreme brevity. The Transistor and the later Integrated Circuit devices offer two tremendous advantages over thermionic valves: they consume very little power and are extremely reliable. They are also extremely compact and the modern computer simply could not have happened without them. Thermionic valves are too bulky, consume too much power and are too unreliable. The marriage between semiconductor devices and computers was made in heaven.

Eckert and Mauchly founded their own computer firm and built their first UNIVAC machine for the US Bureau of Census in 1951. The firm was taken over by Remington-Rand and continued to build a range of UNIVAC computers. Initially most sales were made to the military, but gradually commercial firms became equally important customers.

In 1953 IBM brought its 701 computer onto the market. The IBM marketing strategy was to lease the machines, rather than sell them. In 1953-54, 17 such computers were leased to military agencies or aircraft manufacturers.

In the late sixties integrated circuits began to replace transistors in computers. Among the first computers using integrated circuits were the CDC 6600, built by Control Data Corporation, and the IBM 360 series. These machines proved highly successful.

It has been estimated that by 1965 there were 31,000 large mainframe computers in use in the world. But smaller computers were soon to make most of the running. First there was a range of what might be termed intermediate computers, produced by the Digital Equipment Corporation, founded in 1957 by Ken Olsen. Their first PDP (Programmed Data Processor) machines came on the market in 1960. In 1963 the PDP 8, a machine using transistors instead of valves, proved highly popular on the market. It cost $18,000, which was cheap for a computer at the time, and was the size of a filing cabinet, which was compact at the time. This was only the beginning. Prices and sizes dropped rapidly with the introduction of integrated circuits and so-called mini-computers that cost as little as $8,000 in 1969. By 1971 there were seventy-five firms making mini-computers.

---

[19] Ernest Braun and Stuart Macdonald, 1982, Revolution in Miniature, 2nd ed., Cambridge University Press.

In the early 1970s IBM replaced the 360 series by the 370 series, in which both the logic circuits and the memory consisted of integrated circuits (ICs). The IC memory constituted a major breakthrough, replacing as it did a variety of more or less clumsy memory devices, such as magnetic drums.

The era of the personal computer, also variously known as desktop-, home-, or micro-computer, was yet to come. Two of the pioneers of these now so ubiquitous machines were Steve Wozniak and Steve Jobs who designed and built the Apple I computer in makeshift facilities in 1975. It used an IC central processing unit (CPU) that cost only $20 and 4kbytes of random access memory (4k RAM). The product was a little rough and sold at $666 only to enthusiasts. Some 175 units were sold in 1976. Early in 1978 they raised some capital and built a more sophisticated unit, the Apple II. By the end of 1980, Apple had become a public stock company valued at $1.2 billion! In 1981 IBM jumped on the bandwagon and brought out its own Personal Computer. They sold 35,000 of them in the first year of production and by 1982 there were just under a million computers in American homes. Ten years later it was 31 million. Several manufacturers came and went and the market settled down to the IBM PC that was now assembled from readily available components by a large number of so-called no-name firms. It was almost like the automobile in the early days, except that the PC's, unlike early automobiles, were all almost identical and were assembled from standard components supplied by the electronics industry. By the end of the century there were still some no-name computers available and many computer dealers assembled machines to personal requirements, but the market was dominated by a few large manufacturers such as Hewlett Packard, IBM, Apple, Dell, Compaq, Fujitsu Siemens, and several more.[20]By 2009 the desk top computer and its even more compact equivalent, the notebook, have become ubiquitous. The usual process of concentration of manufacturers has taken place; there are not many left. On the other hand, a large number of computer service people assemble computers to order from standard components readily available on the market.

Prices have not really come down much over the years; instead specifications have gone up by leaps and bounds. The perfectly ordinary computer I am writing this on has specifications that early mainframe computers could not even have dreamed of. A microprocessor that operates at a rate of 750 MHz, 64 MB RAM, 20 GB hard drive, DVD ROM, CD reader-writer, 56 kbit modem[21], and a whole lot of software, not to mention a printer and a flat-bed scanner. As almost all private users, I need only a fraction of these capabilities. Indeed many owners of home computers use them for games and/or as sophisticated typewriters only. Some use a spreadsheet, a database, graphics and e-mail and some further Internet facilities. What most people would like, and cannot get, is reliability – crashes are all too frequent – and freedom from vicious virus attacks. In general, one suspects that most of these machines stand idle for much of the time and many users are irritated by the unpleasant habit of the machines to do automatically what nobody had asked them to do. They display a kind of wilfulness considered helpful by the pogrammers and irritating by the users. Unfortunately the home computer, despite its undoubted advantages, has brought its own problems of crime. Computer crime is of two kinds: either involving plain fraud or theft; or the misuse of chat-rooms, especially those used by children. The Internet has proved a willing and able carrier of advertising, most of it unwanted by the user. Much to the outrage of the moralist, it has also proved a useful medium for pornography. On the other hand, there is quite a lot of useful information on the Internet and this is used by private citizens, business, and students at all levels, though it is not as easy to use as the less skilled user would like. Some of the demand for personal computers is driven by schools and their students.

In 1980, the IBM PC needed an operating system. By some chance coincidences, the choice of system supplier fell upon the small firm Microsoft, founded by the young entrepreneurial software wizard Bill Gates. By 1986 Microsoft had become a public company and by the end of the century Microsoft had become a giant company. Bill Gates had become one of the richest men in the world and a succession of lawsuits, alleging abuse of a monopoly position, dogged Microsoft. New versions of operating systems were introduced at frequent in-

---

20 Apologies to the manufacturers not named here. I am sure they will understand that the names given are just examples chosen at random.

21 By the time of final revision of the text, in early 2009, many of the above specifications have become obsolete. I now use a notebook with much better specifications and a broadband connection to the Internet. And I still am nowhere near using all this sophistication.

tervals and numerous customers complained about system faults. Instead of gradually ironing out the faults, new versions with new faults succeeded each other in rapid succession. It is easy to see that this policy makes commercial sense, but it is equally easy to see that it is irritating to the customer. New systems mean new sales, the painstaking improvement of old systems brings customer satisfaction. Commercial firms rate sales more highly than satisfaction. The customer can, theoretically, choose a different system. In reality, however, the very fact that such a vast majority of operating systems is of the Microsoft PC type makes a change very difficult because of compatibility problems and because of problems of availability of application programmes compatible with other systems. Whether Microsoft is guilty of monopoly machinations or not, the Microsoft systems have a de-facto monopoly position with only a small band of enthusiasts choosing Apple computers, with their own system, instead.

This is all I have to say about stand-alone computers. The rest of the computer story belongs to networking. It all began in the USA with classified military and national security computer nets in the 1970s. In 1983 a distinct military computer net, the MILNET, became established that later merged into the Defence Data Network. At about the same time university computer departments established their own net, the Computer Science Research Network. In 1985 the National Science Foundation (NSF) agreed to build and manage a net connecting its own five main computer centres, the NSFNET. The NSF agreed that their network could be used by commercial users for on-line computer services. In a parallel development, the commercial firm CompuServe, part-owned by Time Warner, started its own network that by 1994 had grown into a huge enterprise with 3.2 million subscribers in 120 countries. Tim Berners-Lee, a computer scientist working for the European Nuclear Research Centre in Geneva (CERN) created protocols for the World-Wide-Web that opened for business in 1992. The NSF withdrew and WWW became a purely commercial enterprise in 1995. Of the many services provided by the net, e-mail is proving the most successful. The almost instant transmission of informal written messages is proving vastly popular. The Internet has become fashionable; most people would not be seen without it. Almost all firms and government departments have web sites. Whether it is for better or for worse, or whether it makes little real difference, who is to say? The fact is that the Internet is not very secure and that its users have great difficulty maintaining the integrity of their systems and their transactions.

The very advances of the personal computer have displaced the mainframe computer to applications where very large amounts of data have to be manipulated. Large government or private administrative offices or large technical or scientific users need mainframe computers; others use personal computers on every desk, networked so that they can cooperate and the network of small machines replaces a larger one. For some highly specialized applications, such as weather forecasting or design of complex engineering systems, so-called super-computers are now used. These are very expensive machines made by very few manufacturers and used in a few specialist centres. The super-computer can save a lot of trial and error experiments in engineering design with mathematical simulation substituting for physical experiments. Design work and technical drawing are now almost invariably computer-aided. This saves a lot of time in producing engineering drawings and it also enables designers in different locations to cooperate on a design by linking their computers.

The home computer has not caused a social revolution. I agree with Brian Winston when he says: "The gap between the hype of revolution and the reality of the underused, complex and extremely expensive consumer durable sitting in millions of middle-class Western homes grew wider."[22] Writing on a computer instead of a typewriter increases the number of revisions and corrections; though whether the quality of the texts increases in proportion must be doubted. The availability of encyclopaedias and dictionaries and many other informative programmes is helpful; the availability of violent addictive computer games is not. Many computer enthusiasts predicted that working from home would reduce employment in offices significantly, to the point that rush-hour traffic in cities would decrease noticeably. No such thing has happened; cities have become even more congested despite the fact that some administrative and professional work is now carried out from home. The impact is not massive, because the amount of such work is not massive. The need for social coherence among colleagues and the need for consultation and supervision are too great for the office to be replaced by a network of lonely people sitting in front of their computer screens. Home shopping, home banking and booking of

[22] Brian Winstone, (1998). Media Technology and Society. London: Routledge, p.238

travel or entertainment over the Internet are perhaps significant, but so far they have not amounted to a revolution, even though it has become difficult to book a cheap airline ticket other than on the internet. For one thing, financial and commercial operations are held back by fear of fraud. For another, people like to deal with real people rather than with awkward-to-use inflexible web sites and they like to handle real goods rather than just see pictures on a screen. People also like to ask their own questions, rather than be fed on a diet of "the most frequently asked questions". No doubt the use of home computers will continue to spread, but many people will be frustrated by the dumb machines replacing real people. For the new generation, however, computers and their mode of transactions are becoming the norm.

The impact of computers on the world of work and on the economy is quite another matter. The Internet is certainly an important contributing factor in the globalisation of the economy. Though it is true that large international firms had their own communications networks even before the Internet, it is nevertheless true that by linking everybody to everybody the global economic network has become more universal and more effective. Administrative procedures have changed beyond recognition. Whereas in the past any change in taxation was a laborious and lengthy process, now governments can easily implement such changes and do so with great frequency – for better or for worse. Other administrative procedures have become automated and impersonal and mistakes, though rare, are not easily spotted or rectified. Telephone information systems are now all on computer and a customer seeking information on almost anything is frustrated by long waits at so-called call centres. Often he/she is confronted by the electronic voice of a computer only, and it takes an even longer wait to reach a human operator. Once reached, the operator knows only what he or she can read off the screen of the computer sitting in front of him/her. The operator has little knowledge and less flexibility. To reach a human with some knowledge and the discretion to be flexible is quite difficult. To reach the person you wish to speak to is often impossible. The computer can only answer questions anticipated by the programmer and, generally speaking, these are not the questions you wish to ask.

It was expected that computers would increase productivity by leaps and bounds. Whether they have actually done so is a moot point. First, the productivity of an administration is impossible to measure because the expected outputs are not quantifiable and can be varied at will. The effort involved in the calculation of a payroll may be quantifiable, but the aftercare of customers, or the services provided for the community, are not. Frantic efforts are often made by governments and managers to quantify the unquantifiable, but all that these efforts lead to is distortion, fraud, and misinformation. In manufacturing the computer probably has increased efficiency by streamlining and optimizing production lines and by linking production to sales. As the name implies, the computer plays an important role in computer-aided design. But industrial efficiency in the wealthy countries is extremely high anyway and industrial products are, consequently, relatively cheap. What the wealthy countries need is cheaper and better service and repair provisions for their industrial goods in order to get away from the environmentally and socially damaging throwaway attitudes. What these societies also need is better provision for public health-care, for care for the elderly, and support for the underprivileged. What societies need is better crime prevention and better prevention of drug abuse and all the other social ills of so-called post-industrial societies. And they need better education in the fullest sense of the word. In all these tasks the computer may have its role to play but it cannot replace humans who think and feel for themselves.

We have discussed several of the defining technologies of the 20th century: automobiles, aircraft, electricity, telecommunications and computers. It is obviously impossible to cover all technologies that characterize the 20th century – there are so many of them. All we can do is mention one or two more and try to summarize what all these technologies have in common and how they contributed to the character of 20th and 21st century society.

There are several more industries that characterise the 20th century. The most important among them are the chemical and the pharmaceutical industries. We cannot possibly describe these in any detail without exceeding by far the limitations of this book. All we can do is describe very briefly the rise of the polymer industry, as the so-called plastic materials are so characteristic for the 20th century.

Polymers

Polymers essentially consist of long chains of molecules that form macromolecules. The properties of the polymer depend both on its chemical composition and on the structure of the macromolecules. Many polymers occur in nature, including living organisms. The original macromolecular theory of polymers was propounded in the 1920s and 30s by the German chemist Herman Staudinger, who received the Nobel Prize in Chemistry for this work in 1953. He laid the theoretical foundation for the plastics industry.

Although some polymers were known and in production before World War II, the great expansion in both range and quantity came in the post war period. As early as 1954, the annual tonnage of plastics manufactured became comparable to the tonnage of aluminium and copper.

Among the earliest pioneers of artificial polymers was Christian Frederick Schönbein in Basle, who discovered cellulose nitrate in 1846, produced by a reaction of paper with a mixture of nitric and sulphuric acid. The substance was suitable for the production of various vessels. In 1869 John Wesley Hyatt produced Celluloid that found applications in the production of billiard balls, collars and cuffs, dentures and more besides. Celluloid film enabled the development of cinematography, although it was eventually displaced by the less flammable cellulose triacetate, first made in 1865 by P. Schutzenberger. C. F. Cross and E. J. Bevan patented an improved method of production in 1894, and it came into large-scale production with the introduction of a cheaper solvent. Cellulose acetate was used during World War I for the fabric of aircraft wings. In 1921 it came into use as acetate rayon and this led to a secondary development of dyes made specially for dyeing rayon fabrics.

The Belgian American Leo Hendrik Baekeland held a series of patents, from about 1909, for thermosetting polymers, known (for obvious reasons) as Bakelite. These polymers are made from phenol and formaldehyde and are generically known as phenolic resins. They are used in very large quantities.

One of the earliest and most important industrial polymers is Polyvinyl chloride (PVC), consisting of long chains of the monomer vinyl chloride (CH_2CHCl). Its discovery is variously attributed to the German chemists E. Simon in 1839 or Eugen Baumann in 1872. PVC was originally extremely rigid. It became useful only in 1926, when W. L. Semon of the B. F. Goodrich Company improved its plasticity in a serendipitous discovery by treatment with some solvents. Another process for plasticizing PVC was obtained by the Union Carbide Corporation in 1930, by copolymerization of vinyl chloride with vinyl acetate. Under the trade name Vinylite it became the standard material for the now extant long-playing record. PVC is used extensively in the construction of domestic appliances and for floor tiles, packaging, food containers, toys, thermal insulation and much else.

Elastic polymers, obtained by the German chemical conglomerate I. G. Farben in the 1930s, played a vital role in Germany as synthetic rubber during the Second World War, when imports of natural rubber ceased. The synthetic rubbers, called Buna, were obtained by copolymerization of two monomers in the presence of a catalyst. The name Buna is derived from one of the polymers, butadiene and the catalyst sodium (Na). The second most commonly used polymer to form a copolymer with butadiene is styrene and the resultant synthetic rubber is also known as Styrene-butadiene rubber (SBR). Synthetic rubbers made a vital contribution to the German war effort in World War II, as it was essential for the production of tyres and other vital goods.

Another important class of polymers are the acrylic polymers, such as polymethyl methacrylate produced by ICI under the name Perspex from 1934. It was used extensively for aircraft windows during World War II, and is now used for a variety of purposes. Early work on this class of polymers dates to 1877 and the first commercial product, polymethyl acrylate, was produced by Rohm & Haas in 1927.

Polyethylene and polypropylene date to the mid- to late thirties, with ICI being among the pioneers. These polymers can be produced in various forms and are used mainly for electrical insulation.

In the United States, E. I. Du Pont de Nemours played a pioneering role in the development of industrial polymers. In particular, the class of polymers known as polyamides were largely developed by Du Pont. Neoprene, another rubber-like substance (elastomer), was discovered in 1932 by one of their researchers, W. H. Carothers. Another famous polymer of the polyamide group is Nylon, announced in 1938 as Nylon 66. The development of Nylon took 4 years and employed 230 chemists and engineers. The development cost was $27 million. The first nylon stockings were produced in 1939 and became an instant success as a replacement for silk stockings. By 1962 there were four different types of Nylon on the market, used for different purposes.

As a description of the history of plastics the above brief account is clearly inadequate, but it will serve my purpose of demonstrating how the nature of technological progress changed even between the 19th and 20th centuries, let alone between the 18th and 20th centuries. Whereas in the earlier period the sole inventor, the practical experimenter, the craftsman and the entrepreneur carried innovation on their shoulders, the later period is characterised by technological progress being made only when the necessary scientific foundations had been laid and the technologists and inventors themselves were academically trained engineers or scientists. More often than not inventions and innovations required very large effort in testing and developing, and a lot of expensive scientific equipment to facilitate this work. Obviously, such large financial costs could only be carried by large firms in the hope of reaping large profits. The development of plastics demonstrates these changes particularly clearly. The days of the lone inventor are largely gone, technological innovation is now firmly in the hands of industrial laboratories.

In the 140 years up to 1930, a total of 4239 patents were awarded for inventions relating to plastics. Of these, 43% were awarded to individuals and 57% to firms. In the years 1946 to 1955 a total of 6238 patents were awarded in the field of plastics and of these only 8% went to individuals and 92% to firms. 36% of all these patents went to the eight most important firms in the field. The example of nylon shows the enormous R&D effort required for a major innovation and shows that although leadership is important, it is the research team that achieves innovations and not the individual researcher. If something like 1,000 person-years has to be invested into R&D to achieve a certain innovation, the likelihood of an individual achieving the same result is remote indeed.

R&D has become a highly organised activity demanding a great deal of money. Some of the money comes from state sponsorships provided under miscellaneous headings and a lot of it comes from private firms, mostly large ones and mostly concentrated in a few high-technology industries. The R&D is usually carried out in large industrial research laboratories, but some of it in public R&D facilities. The universities still play a role in technological innovation, mostly through their efforts in pure research, but also in applied research, often carried out in collaboration with industrial firms.

There is no end to the range of products made from plastics, often replacing other materials such as metals, wood, leather, or glass. Virtually all kitchen utensils are now made of plastics and plastics are used in the construction of domestic machinery, in automobiles, in building and in furniture. Plastics reinforced with carbon or glass fibres are now used in the construction of parts that need to be very strong and light, including sections of aircraft. Sports equipment uses similar materials, whether in the construction of skis or of boats and masts. Virtually all weatherproof clothing is made of plastics and plastics play a large role in all other clothing, footwear, household linen and so forth. Were it not the age of information technology, the 20th century might well be called the plastics age.

Epilogue

The 20th century has seen technology move centre stage. This is true in several respects. First, technological innovation has become recognized as the prime factor in economic growth and in the competitive position of firms and countries. Firms have to produce innovations in order to stimulate flagging saturated markets with offers of new products. They have to use improved methods of production in order to compete on costs, but sometimes also on quality and reliability. New production methods are usually designed to save labour, but automated machinery often also works more accurately and with fewer errors than human operators. The result of all this is that industry employs a diminishing proportion of the labour force. A secondary result is that industrial products are now relatively cheap and consequently they get thrown away when they malfunction, because the cost of repairs is relatively high and competent repairers are scarce. From the environmental point of view this is deplorable, as it wastes energy and raw materials, unless the products that are thrown away are easily recycled.

Governments do all they can to stimulate technological innovation. They support much R&D directly, both military and civilian. They also try to introduce all kinds of incentive schemes to promote R&D, whether in terms of tax relief, or in government laboratories, or in attempts to foster R&D cooperation at the national or international level. The European Union is quite active in this field and supports many R&D projects, always run on the basis of cooperation between firms in different countries and often including universities. The proportion of national income spent on R&D, both private and public, is a matter of national pride and governments try to push it up year by year.

Because a high proportion of R&D is always spent on defence projects, governments attempt to get as much civilian benefit out of these projects as possible. Dual use is one of the favourite slogans of the day, meaning technologies that are useful for both military and civilian use.

It has become customary to speak separately of those industries that require high R&D expenditure for survival and call them "High Tech Industries". They include electronics, computers, aerospace, and pharmaceuticals. There is a second tier of industries, including the automobile industry and the chemical industry, that requires substantial R&D effort, though not quite as much as the first group. And then there is the rest of industry, often disparagingly called the old industries, such as textiles, food and drink, bulk chemicals, iron and steel, heavy electrical goods, machine tools, building, and so forth. The division is not very helpful, as the older industries still supply the bulk of our needs, though the so-called sunrise industries supply all the fickle industrial glamour. Industries from high wage countries have transferred many labour-intensive operations to low-wage countries and newly industrialising countries have often concentrated on the older industries because they demand less know-how and are more suitable as measures for import substitution. Textiles and shoes, for example, are now produced largely in cheap labour countries.

The second sense in which technology has moved centre stage is the fact that we are surrounded by it everywhere and all the time. On the street, in the office, in the home, even on holiday we use technology wherever we go and whatever we do. The office is now full of computers, photocopiers, telephones, fax machines (already declining after a very short life), coffee machines, air conditioning, and sophisticated lighting. The street is full (very full) of motorcars, other motor vehicles, traffic lights, mobile phones, and personal tape and CD players. The home is stuffed full of machinery and equipment that did not exist at the beginning of the century; from washing machine to television set and from home freezer to home computer. Domestic machinery has had a big impact on the possibility of freeing the housewife for outside employment, though I do not wish to argue either way whether technology freed the housewife or the free housewife demanded the technology. Perhaps a bit of each.

Medicine has certainly learned to cure many ills, but the technological advances have caused severe financial problems in the provision of medical care for the population at large. People live longer and diagnostic, surgical and therapeutic equipment have become sophisticated and vastly expensive. To make matters worse, technical progress and commercial acumen of the equipment manufacturers are causing rapid obsolescence of medical equipment, so that replacement purchases add to the burden of new purchases. Some argue that the attempts to keep hopelessly ill and suffering patients alive beyond their wishes adds unnecessarily to costs and to the sum total of human suffering. This is undoubtedly true, but the ethical problems that need to be solved to cure this ill are formidable, yet not insoluble. Some tentative steps are being made to take the will of severely sick patients into account when decisions on keeping life-support machinery going or not are being taken.

Advances in travel and in communications have contributed to shrinking the world, for better or for worse. We know more about the world, we see more of the world, but we have to put up with globalization of the economy that is tantamount to our lives being dominated by major global companies. Even in democracies the influence of major industrial players is highly significant.

Military technology has advanced to almost unbelievable levels. Aircraft are incredibly fast and efficient. Modern artillery and its munitions are highly efficient. A whole range of different missiles supplement or supplant artillery. We have surface to air missiles, surface-to-surface missiles, sea to surface missiles and all other combinations of missiles with all ranges and all guidance systems. We have huge aircraft carriers that are virtually floating air bases, escorted by a flotilla of ships with incredible firepower. The modern military engineer has bridges that can be assembled in no time, so that rivers no longer form major natural obstacles. Fortifications and bunkers stand little or no chance against modern bunker-busting bombs. Soldiers in trenches stand no chance against modern bombers. Older tanks stand no chance against their modern counterparts or attack aircraft.

The problems of environmental destruction have changed their nature but have not become less worrying. Whereas at the beginning of the century gross pollution of the rivers by industrial effluent and gross air pollution from burning coal were the dominant problems, we now have made great efforts and more or less cleaned up the rivers, though occasionally chemical spills still kill thousand of fish and insidious pollution persists and causes much harm. We have got our act together and cleared the air from gross pollution, only to replace it by more insidious pollution from car exhausts and other sources that cause the highly worrying greenhouse effect and associated climate change, as well as the depletion of the ozone layer and micro-particles in the air. We still kill our birds by pesticides and fungicides and who knows what else. We are depleting our resources in raw materials and in oil and gas. We are destroying our remaining rain forests, destroying thousands of species of flora and fauna and removing an important part of our source of oxygen. Shall we destroy our habitat completely?

Technology has made us infinitely wealthier. The last hundred years have seen an unprecedented growth of wealth. It has also seen an unprecedented growth in the variety of goods and services available to us. We consume far beyond our needs. Both the supplier and the consumer have become greedy. Yet many people, even in the rich countries, live in relative poverty and not a few in absolute poverty. The poor countries are as poor as ever. Not many people in the advanced countries do physically hard work, but most people have very strenuous jobs and lead fairly stressful lives. What is the balance? Has technology made us happy? If we had a time machine, would we go back a hundred years? Or fifty years? Or not at all?

Technology has become closely associated with science. In fact various branches of physics, chemistry, biology, and mathematics have been integrated into branches of engineering. Applied mathematics, solid state physics, genetics, organic chemistry, to name but a few, operate in a close symbiosis with engineering. It is no exaggeration to say that engineering and science have become a joint activity, with some aspects of each being more practical and applied and some more theoretical and pure. The marriage of science and technology, consummated in the 20th century, appears to be stable and fruitful.

Main Literary Sources Used for this Chapter

Angelucci, Enzo and Alberto Bellucci. (1975). The Automobile: from steam to gasoline. London: Macdonald & Jane's.
Baldwin, Neil. (2001). Edison: inventing the century. Chicago: University of Chicago Press.
Brandon, Ruth. (2002). Automobile – How the Car Changed Life. Basingstoke: Macmillan.
Braun, Ernest. (1984). Wayward Technology. London: Frances Pinter Publishers.
Braun, Ernest and Stuart Macdonald (second edition 1982). Revolution in Miniature – the history and impact of semiconductor electronics. Cambridge: Cambridge University Press.
Davis, Martin. (2001). Engines of Logic: Mathematicians and the Origin of the Computer. New York: W. W. Norton & Co.
Eden, Paul and Soph Moenk. (2002). Modern Military Aircraft Anatomy. Leicester: Silverdale Books.
Endres, Günter ed. (1998) Modern Commercial Aircraft. London: Salamander Books.
Gay, Peter. (1977). The Enlightenment: an Interpretation. The Science of Freedom. London: Norton Paperback.
Gordon, John Steele. (2002). A Thread Across the Ocean: the heroic story of the transatlantic cable. London: Simon & Schuster.
Hughes, Edward (1995, 7th edition, revised byI McKenzie Smith). Electrical Technology. Harlow: Longman.
Jenkins, Lawrence. (1987). Digital Computer Principles. New York: John Wiley & Sons.
Nader, Ralph. (1965) Unsafe at any Speed. New York:Grossman Publishers Inc.
Pool, Robert. (1997). Beyond Engineering – How society Shapes technology. New York: Oxford University Press.
Porter, Roy. (2001 2nd ed.). The Enlightenment. Basingstoke: Palgrave.
Pucher, John and Christian Lefèvre. (1996). The Urban Transport Crisis in Europe and North America. Basingstoke: Macmillan Press.
Singer, Charles and E. J. Holmyard and A. R. Hall and Trevor I. Williams (Ed.). (1956). A History of Technology, vol. V, The Late Nineteenth Century. Oxford: Clarendon Press.
Wangensteen, Owen H. and Sarah D. Wangensteen. (1978). The Rise of Surgery – from empiric craft to scientific discipline. Folkestone: Dawson.
Williams, Trevor I. (ed.) (1978). A History of Technology. Volumes 6 and 7, the twentieth century: Oxford, Clarendon Press.
Winner, Langdon. (1977). Autonomous Technology – Technics-out-of-Control as a Theme in Political Thought. Cambridge, Mass.: MIT Press.
Winston, Brian: (1998). Media Technology and Society. London: Routledge.
Womack, J. P. and D. T. Jones and D. Roos. (1990). The Machine That Changed The World. New York: Rawson Associates.

Chapter 7

Technology and Society

Considering that the title of this book is *From Need to Greed*, we shall define these concepts first, even at the risk of repetition. Need is the requirement for the availability of conditions and technologies that enable human life to exist at a basic subsistence level. Needs are satisfied by basic essentials, such as basic food, shelter, clothing and clean water. According to circumstances, the list may be extended to include cooking facilities and some kind of heating in cold climates. If we wish to be slightly more generous, we may include a requirement for basic education and health care. What we regard as basic varies with social development. What is basic now was extremely luxurious, indeed non-existent, in earlier times.

Greed, at the other extreme of the scale, drives our desire for possessions and material satisfactions well beyond what is necessary for survival. Purchases of luxuries are driven by greed. Thus purchases of expensive jewellery, meals in expensive restaurants, holidays in luxury hotels, ownership of yachts and luxury cars or private jets are all driven by greed. The desire for personal prestige and power can also be partially satisfied by purchases of luxury goods.

In the continuum of gradations between need and greed we may introduce the concept of want, which describes our desire to live comfortably above subsistence level and to indulge in a modicum of small luxuries. All three concepts, need, greed and want, are largely socially determined and vary from time to time and from place to place. In very rough general terms, wants are satisfied by the purchases of the middle classes that extend beyond the purchases of the poor, but fall behind the purchases of the rich. Obviously these definitions are purely indicative and not rigorous.

We distinguish between technologies that are driven by the need to survive and serve only our basic needs, and technologies that are driven by the desire for comfort and fun. Hence we say that although technology still satisfies our needs, its main focus has shifted to satisfying our greed and our wants. In order to survive as a producer of technology in a capitalist society, profits are of the essence and thus the term greed is here also used in a sense that describes a desire for profit. For lack of a better term we thus use the word greed with two slightly different meanings; the desire for excessive consumption and the desire for profit. The assertion made in this book that technology has gone well beyond fulfilling human needs, has two components. Producers of technology no longer answer the question what do people need; instead they ask what technology they can profitably sell to people. People, on the other hand, no longer buy only what they need in the strict sense of the word but buy to satisfy their wants or their desire for luxuries. The title from need to greed thus describes a fundamental change in behaviour of both producers and consumers of technology.

We next define society in the modern sense as an assembly of people living in proximity to each other in both space and time. People living close together inevitably interact with each other and need rules to avoid or regulate conflicts. They also need to organise the supply of goods and services they require both for survival and for pleasure. Quite apart from rules and organisation, individuals are strongly affected in their behaviour and their thinking by their interaction with other members of their society. Over the millennia of human existence, many forms of societies have evolved and indeed societies are continually evolving. Social organisation is complex and multi-faceted, possessing the following features in various forms:

1. A legal framework that is vital for the avoidance of conflict and, even more to the point, for not allowing conflicts to get out of hand. Thus we have rules of what is allowed or forbidden, rules how certain social trans-

actions are to be carried out and we have a system for adjudication in cases of conflict and for meting out punishment in cases of violation of the rules. The legal system consists of many sub-systems. In democracies, parliaments make the more important laws and many lesser organisations make lesser laws and rules. There is a bureaucracy to implement the laws and regulations, and to provide the infrastructure a modern society requires. We may distinguish between a social infrastructure, e.g. regulations and public administration, and a technological infrastructure.

2. Underpinning the society is a multi-layered educational system that provides educated citizens as carriers of the elusive system of customs, beliefs, assumptions and traditions which, taken together, define the characteristic features of a given society, often called its culture. Unfortunately the word culture is somewhat ambiguous, especially as the word describes the general ambience of a society but also its cultural activities, such as art, literature, music, science. I like to divide, somewhat controversially, this aspect of culture into high culture and popular culture and thus I distinguish, for example, popular music from classical music. Culture is an ambiguous concept that can be properly understood only in the context in which it is used. The educational system is also called upon to provide skilled people for work on the infinite number of tasks that need to be carried out in an advanced society, from bricklayer and carpenter to architect and structural engineer; from general medical practitioner to specialist surgeon, from school teacher to nursery nurse, from actor to musician and so forth to almost infinity.

3. For a society to function on a large scale it needs a substantial technical infrastructure. This consists of a system of transport and communications, a system of housing, a system of manufacturing all the multitude of artefacts that society requires (whether it needs them or not), a system of food production, a system of maintaining public order and a system of waging war (also known as defence). Of course all these systems require an elaborate organisation and, what is of the greatest interest to us, a huge variety of artefacts known as technological products and a huge variety of techniques for the manufacture of such products and for the maintenance and construction of the technological infrastructure.

As society evolves, so does its technology. We have argued elsewhere[1] that whereas technology develops unidirectionally, social structures and, first and foremost, the values held by society evolve in a haphazard, somewhat cyclical, manner. At the time of writing this last chapter of *From Need to Greed*, in the spring of 2009, there is very little cause for rejoicing when surveying the social situation of large parts of global society. We have a combination of unrestrained capitalism that has driven itself into a catastrophic financial and economic crisis, causing much hardship and pain. At the same time we have severe environmental problems that are being tackled rather inadequately, allowing a serious degradation of the planet. On the other hand, we have vicious dictatorships and bloody strife in many countries. It is, of course, always possible to remain sanguine as to the future, but it is very hard to see the world in its current state as a just, fair and wonderful place. Nature is undisputedly wonderful, but human greed is busy destroying what remains of it. Surveying the state of human society and the state of the planet on a global scale does not warrant complacency.

This book deals with the role of technology in society over a long period of history. One of the most characteristic features of technology is its constant development and this means that the consideration of technological innovation is uppermost in our story. We attempt, however, to consider technological innovation not only as the outcome of the inner logic of technology, but also as the result of social forces influencing technological development. We have to think of technological change in a changing society. Technologists and their managers are permanently on the lookout for commercial opportunities. To say that technologies are developed only in response to needs is a smokescreen; the reality is that needs are developed in order to make new technologies profitable. The commercial principle that guides technology is the seeking out of opportunities for making money with the aid of technology. Technology is the most fundamental means for making profits, because technology is the means for making goods that can be sold at a profit. There are many ways of making money, but technology is the only one that also creates new real tangible material wealth.

Of course technology is still required to fulfil its original purpose; to sustain human life on earth. But whereas in the past technology was only able to provide a small population with the most essential needs, ena-

[1] 1995, Ernest Braun, *Futile Progress*, Earthscan, London

bling them to live short lives full of toil and hardship, modern technology sustains huge populations at varying levels of comfort. Whereas life in developed countries has, by and large, become extremely comfortable, nay luxurious, and the life-span now exceeds the biblical three score and ten, life for the majority of humans who live in less developed countries is still pretty hard and miserable. Life in modern Western cities requires an extremely high level of technology just to keep going. A modern city needs vast amounts of food to be distributed, it needs all forms of mechanical transport, water supplies, sewerage, electricity, heating, air conditioning, medical services, security services, and much else besides. Instead of making hopeless attempts at enumerating all the technologies required to sustain life in a city, we may subsume them all under the title *system of city technologies*. At a basic level, these technologies may justly claim that they provide only what people need, as without them life in cities would simply collapse.

Finding new uses for various materials can also create new wealth and economic growth. Metal ores had no value until metal smelting and the use of metals for the production of various artefacts were discovered. The value of precious stones or precious metals is not inherent in them – their value derives from their fashionable use in highly valued artefacts such as jewellery, but also in the value ascribed to them by rarity and convention. Finding new uses for materials or introducing new items from foreign lands is, as far as the economy and economic growth are concerned, as good as new technologies. A good example of the extreme value of a new use for a material is the case of oil. Oil was originally just dirty useless stuff. It began increasing in value when it became used as raw material for the production of various substances and its value truly exploded when motorcars, power stations, domestic heating systems and industrial machinery – to name but a few - developed an insatiable thirst for the stuff. Oil has become one of the most important trading items in the world and vast commercial empires are sustained by it. The process can, and does, also happen in reverse. When flint was no longer needed for the production of tools and weapons, it lost whatever economic or traditional value it had possessed.

Technological innovation creates new commercial opportunities and is therefore regarded as highly desirable. Because markets for technological products tend to saturate, technologists and their managers try to rekindle flagging sales by improving, changing, or replacing products when sales stagnate. Apart from rekindling saturating markets, technological innovation creates entirely new products that create entirely new markets. From the point of view of the economy a new service is almost as good as a new product, as long as a reasonable balance is maintained over a range of services and products. Generally speaking, new services are often connected with the use of new technologies. The introduction of the computer, for example, has brought in its wake a number of new services, such as computer programming, systems analysis, computer repairs, computer instruction, and so forth. The computer has also fundamentally changed other services, e.g. accountancy, banking, tax collecting and many more. Technologists and investors seek to find applications for new technologies wherever they might find them and it matters little to them whether they make a profit out of selling new products, or new production technologies, or new services.

Until quite recently services such as retailing used only very simple technologies. Weighing-scale and cash register were just about all the technology needed by retailers. The IT revolution changed all that. Self-service has largely replaced shop assistants. The goods are automatically pre-packed and labelled with bar codes and thus carry their own information on their weight, the unit price and the price of the package. At the till the label is automatically read and the total price of the goods in the shopping basket is worked out. At the same time the sales information is collected for purposes of stock-keeping and as a basis for the formulation of sales policy.

Many services have arisen out of the application of new technology. Indeed most major new technologies bring in their wake new services needed to use, service, and sell the new technologies. Older technologies, as the example of retail sales has shown, often deploy new technologies when these appear to provide advantages for their particular business. The development of Information Technologies followed its own trajectory and in the course of this development it became obvious that retail sales could use some of the new technologies, whereupon specific developments aimed at the retail trade took place. The sequence of events was that technology developed to a certain point and technologists then became aware of a possible new field of applications. This field was then further developed in conjunction with the newly targeted customers. At least some aspects of IT have managed to conquer a great variety of markets and IT has become a universal tool for a great many human activities.

Numerous writers, considering technological innovation over the last few decades, have aimed mainly at classifying types of innovation and analysing ways and means of organizing innovation so as to accelerate its pace and improve it efficiency. Much of this kind of writing aimed to improve policies supporting technological innovation, but also policies for controlling unwanted effects of new and old technologies.

As an example of the classification of technological innovation we may take the taxonomy described by Freeman and Perez [2], who distinguish between four types of innovation:

1. Incremental innovation, i.e. the process of continuous improvement and development of products and processes;
2. Radical innovations, i.e. technological discontinuities which are usually the result of deliberate R&D;
3. Changes of technology systems, i.e. 'far-reaching changes in technology, affecting several branches of the economy, as well as giving rise to entirely new sectors';
4. Changes in techno-economic paradigm (technology revolutions).

This and similar taxonomies have a certain intellectual beauty, though they classify technological innovations a posteriori only. They are useful for obtaining a historical picture on the role of certain types of technologies in the overall scheme of technological development. However, it needs to be emphasized that items two to four of the above taxonomy do occur over a long period of time and really consist of a whole series of innovations. A single big leap forward is very rare, if it exists at all. Large steps in technological developments usually consist of a series of small steps and both the starting and finishing lines are set somewhat arbitrarily. Take the innovation "steam driven railways". Undoubtedly, an innovation of the highest order, indeed a change in a techno-economic paradigm. The leap from animal drawn carts and carriages, requiring only simple roads and not much organised infrastructure, though the effort of providing roadside inns and stables to facilitate long journeys should not be underestimated. Nevertheless, building railroad track, steam locomotives, carriages, and a supply and maintenance network requires a great deal more and the investments and economics of railways is of a different order to that of animal-drawn transport. But where did this innovation begin and where did it end? There certainly was no single idea or driving force behind it. Nobody sat down and decided to design a new transport system from scratch. What happened was that individual smaller innovations created the conditions that made the steam-driven railway system possible. Many things had to come together. First the steam engine, which itself was created in many small steps. The steam engine had to become light enough and small enough to be mounted on wheels. This "miniaturization" also required a number of small steps taken in order to allow a more widespread use of steam engines rather than specifically with railways in mind. This thought came somewhat later and required several crucial improvements to coal-fired steam boilers. Other innovations that fed into the stream that ended with the early railways were the improved manufacture of steel rails, the construction of stable tracks, the production of strong wheels and so forth. And what is the end point of this innovation? One possible answer is that we have not reached the end point yet. Railways are still improving and still experimenting and innovating, reaching greater speeds and less pollution. The railway connecting London to Paris and Brussels via the channel tunnel is certainly a very far cry from the early railways constructed by Stevenson or Trevithick and even from the tracks built by Brunel. We might agree to set the end-point of the railway innovation at some arbitrary point when the building of track proceeded at pace and railways were run on a commercial scale. Or should we set the end point at the end of World War One, or World War Two? Or perhaps with the activities of Lord Beeching in 1963, who wrote a report that led to radical pruning of the huge British railway network to a more economic size, and thus set the point at which railway development in Britain reached its peak in the total length of track?[3]

[2] Christopher Freeman and Carlota Perez, A taxonomy of innovations, pp. 45–47, in Giovanni Dosi et al., eds., , Technical Change and Economic Theory, 1988, Pinter Publishers, London & New York

[3] Between 1963 and 1975 the length of railway track in Britain was reduced from 28,000 to 17,000km

There are two points I am trying to make. First, the classification of an innovation depends to a considerable extent on the points chosen for the beginning and for the completion of a particular innovation, and both these points are fairly arbitrary. Secondly, most innovations, except the least radical and often trivial ones, require a large number of preconditions and consist of many innovative steps that, taken together, constitute a given innovation. Thus an innovation may be seen as the integral over time of many small contributing developments and many preconditions that are necessary before the innovation can proceed.

Many of the preconditions are technical, i.e. requirements of pre-existing technologies. Many conditions are social and consist essentially of a supply side, i.e. the availability of capital for investment in the innovation and of suitable labour that either has, or can acquire, the skills necessary for the development and application of the new technology. The demand side, as always, consists of the probable total demand for the innovation, in the case of railways this means total demand for freight and passenger transport that might be carried by rail. The demand forecast for an innovation is always a forecast and therefore largely a matter of faith backed by common sense.

We have stressed many times that an innovation is basically a coming together of a new technological possibility with a social demand. In other words, the new technology needs to arouse sufficient interest by a sufficient number of purchasers to part with sufficient cash to make the innovation economically viable. Thus the idea of an innovation will only come to fruition if the new technology meets sufficient demand, backed by purchasing power.

In the distant past technologies were created for the sole purpose of enabling humans to survive. Human inventiveness and human needs and wants led to constant improvements in technology and gradually enabled humans to tackle a greater variety of tasks of increasing complexity. Eventually, through a combination of technological logic and social desires, humans abandoned their nomadic hunting and gathering existence in favour of a settled way of life, supported by agriculture and animal husbandry. Sedentary societies became increasingly complex with the rise of wealth and the division of labour.

When humans embarked upon the path of agricultural production and urban life, gaps appeared in the technological system that supported the new way of life and thus new needs arose that had to be satisfied by new technology and new organisational measures. Large settlements developed in regions of high agricultural productivity and interdependence between agricultural production and urban markets developed. The operation of this system required improvements in agricultural techniques in order to supply a growing number of non-agricultural workers with food. It also required the establishment of a system of transport into the towns and a system of storage of food for a sizeable population. The distribution of food required some form of organisation: it could be traded, which required markets and money, or some kind of tally system, or it could be distributed by some authority, and that necessitated a suitable organisation and a written accounting system. It is a fundamental property of all technological systems that in the course of their development gaps arise that need to be filled by new technologies and/or new organisational measures to guarantee the smooth functioning of the system.

Nomadic hunters and gatherers own very little. The concept of wealth became established as a result of the rise of agriculture and the sedentary lifestyle. The most important early form of wealth was ownership of land, but this was soon supplemented by forms of wealth created by technology. Indeed from very early on technology became a vital factor in the creation of wealth, and the pursuit of wealth soon became a driving force for the creation of new technologies. Technological artefacts became embodiments of wealth. This is particularly true for the ownership of dwellings, from the modest family home to the luxurious palace or public building. Domestic implements, furniture, and ornaments also soon claimed their place amongst objects of value. Whereas the farmer attempts to own as much land, and as many farm animals and farming implements as possible, the urban dweller attempts to own as large a house as possible and furnish this with the best furniture, domestic implements and ornaments.

The most vital change in the human condition, brought about by the introduction of agriculture, was the change from a nomadic to a sedentary way of life and the establishment of permanent settlements, many of which later grew into towns. The rise of agriculture and its consequences are a prime example of the intimate interweaving of social and technological factors. We may say that agriculture was a logical development of technology, and society found it convenient to adapt to its use. But we might also say, with equal justification, that society found it convenient to become sedentary and developed the technology to enable it to achieve this aim. I do not think that this chicken-and-egg argument can be resolved. We have to accept that social and technological factors form an inseparable pair, and that the fabric of society is a weave of human and technological factors that cannot be neatly separated into weft and warp, social and technological.

The second vital change in the human condition, also a consequence of the establishment of permanent settlements, was the emergence of social organisation, including the establishment of religious cults. No doubt early nomadic tribes had some rudimentary social structure, but nothing as complex and tangible as sedentary populations. It is probable that some form of animistic beliefs emerged very early in human existence. The incomprehensibility of death and the many puzzling and frightening aspects of nature probably led humans to believe in spirits – perhaps the spirits of the deceased – that may intervene in human affairs. We cannot know what hunters and gatherers thought, but we do know that Neolithic people built large edifices and complex burial places that provide plausible evidence for organised religion. The large monuments also provide evidence for social organisation, probably cooperation at regional level, for the enormous amount of planning and work that these edifices required could not have been accomplished without leadership of some kind and without labour coming from far and wide. The need for religious or social cult was strong enough to lead to its physical embodiment in the shape of massive works, such as Stonehenge and many similar ancient monuments. Technology began to serve non-material needs that were deemed desirable, in addition to supporting physical survival. It would be wrong to argue that society placed orders for monuments, and technology fulfilled them. The need was primarily social rather than material, and there was interplay between the need and technological capabilities, a kind of bargaining between social and technological forces, which led to decisions on what technology could provide and on the degree that this would satisfy the social need. We do not know the mechanisms by which such bargains were struck, but struck they were.

Toward the end of the Neolithic period technological skills had developed to a degree of complexity that required specialist craftsmen. In earlier days, the people who used the tools also produced them, later on there were specialist producers at work. Some trade in certain high quality raw materials and in technological products had begun to develop quite early on. It was not yet a trade in which the traders accumulated wealth, but a trade that enriched local facilities and caused an exchange of ideas and knowledge over some considerable distances. The precondition for both specialisation and trade was, of course, that agriculture was able to produce more food than the farmers and their families consumed themselves. Agricultural surplus is a prime condition for the existence of non-farming occupations.

As societies became more extensive and more complex, the non-farming occupations that arose included rulers, administrators, priests, specialist technologists, traders and, soon enough, soldiers. The accumulation of wealth by some, and the formalisation of religion, led to social differentiation and to the exercise of power by an elite. The elite was associated or allied with priests; earthly and supernatural powers became firm allies. It is easier to exercise power over fellow humans if the claim to power appears to be backed by divine authority. Relatively conspicuous consumption by the elite enhanced their prestige and their power.

Civilizations, from ancient times to this day, have two faces: the military, conquering, cruel face; and the face of organised society, the rule of law and high artistic achievement. When we speak of civilizations, we should remember the two meanings of the word: on the one hand civilized peaceful behaviour with high artistic and technical achievements, and on the other hand power, warfare and cruelty. War became a normal aspect of the life of states and has remained so to this day. Whereas technology originally served the purpose of physical survival, and thus the maintenance of human life, technology later accepted the task of serving the purposes of war and thus contributes massively to the extinction of life. Ever since the early civilizations, the task of producing weapons and armaments has been one of the major preoccupations of technology, and never more so than in most recent times. Undoubtedly early tribes fought neighbouring tribes to some extent, but truly organised warfare and organised armies belong to a period of major cities.

Once the concept of wealth was established in people's minds, the pursuit of wealth inevitably followed and this pursuit took on many forms, such as trade, conquest, slavery and exploitation of fellow humans. As technology developed, its production became the domain of the skilled craftsman who had several possibilities of acquiring wealth, even if his income often sufficed only for a modest living. The skilled technologist could sell his skilled labour to those who needed it, or he could sell products of his labour in the form of traded goods such as furniture, domestic utensils, tools, carts, or weapons. As soon as the avenue to wealth via technology had been created, technologists began to think of new possibilities for increased revenues. The first thing that comes to mind is, of course, the creation of opportunities for new income by the creation of new technologies. Thus the relationship between technological innovation and the creation of wealth became established quite early on, when technologists were simple craftsmen working in small-scale workshops, long before the establishment of capitalism as a form of social organisation. Farmers, on the other hand, could increase their income either by acquiring more land, or by increasing the efficiency of their production by means of improved technologies.

We may ask what modern society requires of its technology. The requirements may be summarised into five categories.

1. Technology needs to sustain the life of the huge population living on the planet today. This task includes the provision of a basic supply of a multitude of goods, including clean water, food, clothing and shelter, but also the provision and maintenance of the essential physical infrastructure needed by modern societies, e.g. for transport by road, rail, water and air, for communications, public administration, health care and education.
2. Technology is called upon to make a major contribution to economic growth by increasing productivity in the production of goods and the provision of services, and by bringing new technologies onto the market, thus extending the range of goods available for consumption and increasing wealth.
3. Modern technology can and must reduce human drudgery. Modern machines largely eliminate the need for excessively hard physical labour. Beyond that, technology provides physical comfort in comfortable housing, the provision of hot water, heating, air conditioning, and good clothing.
4. Technology needs to provide the material means to safeguard the internal and external security of modern states.
5. Technology needs to find solutions to societal problems as and when they arise. Examples are problems of environmental pollution and climate change, or threats to human health, e.g. by malaria or aids.

The above are the functions society expects its technology to fulfil to a greater or lesser degree. The producers of technology, on the other hand, expect to make profits from technology. They make profits on condition that their goods find buyers willing to pay a price that allows the producer a profit margin.

It is interesting to ask whether technological development obeys some fundamental laws, or rules, and, if so, what these laws might be. We may define six such laws:

1. **As technology develops, it expands its capability for the satisfaction of an increasing range of human wants**. The wants generally increase from essential needs for mere survival towards wants that tend toward the unnecessary and luxurious. Necessity is, however, at least in part determined by social structure and custom.
2. **As a technology develops in the course of its life, its performance, measurable by an individually defined factor of performance, improves**. The steam engine, for example, improved to produce greater power with reduced size and lower fuel consumption.
3. **Technological development is irreversible and always proceeds in the directions given by the first and second laws.**[4] Technological progress means the satisfaction of an increasing range of human wants and an improved performance of technological devices. Change in the opposite direction does not occur. It is arguable, however, that some qualities of products, such as durability and appearance, may deteriorate as a result of progress.

[4] The termination of supersonic passenger flights appears to contradict this law, but Concorde was a politically created blind alley in technological development and the advantage of its great speed was outweighed by many disadvantages.

4. **Technology progresses into technological systems with increasingly complex interdependencies.** The interdependencies occur in the means of production, in the input of parts and in the actual use of the new technologies.
5. **The introduction of new technology often involves painful adjustments.** These occur in society, particularly because of the obsolescence of some skills, the displacement of labour and of some products by others.
6. **No technology that goes against the interests of the ruling elite is ever accepted.** At present, labour saving technologies are readily accepted. Technologies that would increase the labour content of a product are unlikely to be accepted. It needs to be said, however, that sometimes the introduction of a technology has unintended and unforeseen consequences and these may, occasionally, be detrimental to the interests of the ruling elite. It also needs to be said that consequences of a technology are usually equivocal and that there may be some benefits and some disadvantages for all concerned. Nonetheless, the above law is generally valid, though there may be occasional exceptions from it. At present the ruling elites appear to be willing to accept that the preservation of the environment is in their interest and therefore there is hope that so-called green technologies will prevail.

The first and second laws simply state that as technologists and their managers and investors seek opportunities to sell technological devices and products, they must be constantly on the lookout for unsatisfied potential demands. This can be done in two ways: either by offering technologies that fulfil entirely unfulfilled potential wishes, (this leads to the first law), or by making improvements to existing technologies, thus kindling new demands for the improved product (this leads to the second law). The SMS (short message service) for mobile telephones is an example that combines both laws. It represents an entirely new product that satisfies a previously non-existent demand, and it represents an improvement in the performance of mobile 'phones. The strategy of the innovators has worked; SMS is becoming increasingly popular, though nobody could argue that it is vital to the survival of individual humans or the human race.

The third law results from the fact that comfort is addictive. Nobody who has used a washing machine would voluntarily go back to washing by hand. Nobody who has used a mechanical digger would want to go back to pick and shovel. And despite the somewhat problematic nature of computers, not many people who have used a word processor would go back to a typewriter.

The fourth law will be discussed later; suffice it for now to mention a contemporary example. The modern desk-top computer is part of a system, comprising, among other items, the following: semiconductor electronics (so-called chips), other electronic components, visual display units, various memory devices, printers, keyboards, audio systems, system software, applications software, the telephone network, the Internet with its service providers, and maintenance and repair services.

We have discussed some of the momentous social upheavals, caused by the introduction of new technologies, especially during the industrial revolution. Much social change was brought about in more recent times by the introduction of computers and automation. The huge factories with tens of thousands of workers are a thing of the past. This is perhaps why old-fashioned socialism, based mainly on the teaching of Marx, is a thing of the past. The proletariat that Marx had in mind no longer exists, which is not to say that there is a lack of underprivileged people even in the rich countries. The modern workforce in manufacturing industry is very much smaller than in the past and many of the highest skills required have moved away from the shop floor into management. Both these trends illustrate the 5th as well as the 6th law, as management forms part of the ruling elite and much prefers to manage a small workforce rather than a huge army of potentially unruly workers. The moderately skilled worker on the factory floor has largely given way to unskilled operators of automatic machinery on the one hand, and very highly skilled machine setters and maintenance workers. All operations are thoroughly planned and designed and many of the previous skills have been incorporated into computer software.

Among the social effects of doubtful benefit we might mention urban sprawl, the demise of the small shop, and an increase in road casualties as results of motorised transport. We might mention the demise of the small farmer as a result of mechanisation of farming, or the demise of thousands of typists as a result of the introduction of computers in administration. We might mention the predominance of large-scale commercial operations as a result of massive savings in mass production, in mass purchasing and in mass marketing. All these trends result from automation, mechanisation, and computerisation and are supported by political upheavals that support neo-liberal policies and globalisation.

As the centuries went by and the range of available technologies became near infinite, the link between the creation of new technologies and the creation of wealth became ever stronger. Though undoubtedly technology still serves our vital needs and enables humans to survive in vast numbers, the main focus of technological endeavour has shifted away from the satisfaction of needs to the search for new opportunities for the creation of income and wealth. Technological innovation aims to create new sources of income, and thus has become a major source of economic growth. Because economic growth is a coveted goal of almost all contemporary societies, and technological innovation is seen as a principal contributor toward economic growth, technological innovation has become a favoured child of society.

As our real needs can now be more than satisfied in Western societies, but the desire for growing wealth apparently can never be satiated, technological innovation is pursued with undiminished fervour with the goal of making money. The desire to make money is the principal driving force behind all contemporary technological activities. This is not surprising in a capitalist society in which the desire for making money is the principal driving force behind all economic and most other activities. Technology is not exempt from the fundamental principle of capitalist society.

Even necessities can be provided at very different levels. If we look at a modern Western household and speculate how much of its equipment is truly necessary and how much simply makes life more pleasant, we invariably reach the conclusion that the modern household is equipped well beyond its essential needs. Not only do we own more equipment than we need, but much of it is substantially more elaborate than it need be. Yet those who cannot afford modern household equipment feel severely deprived and relatively poor.[5]

Profits and wealth can be made out of a great variety of activities. They are, for example, made out of trade, even though the trader does not produce material goods. Trade is essential for the distribution of goods and, as far as the economy as a whole is concerned, the source of profits is often seen as immaterial, though I regard this view as fallacious. Some argue that as long as money is made and circulated, the economy benefits. However, real material wealth is only created by technology, including agriculture, and no amount of trading could keep an economy going without material goods being produced – though not necessarily in the domestic economy. It is also hard to imagine how economic growth could be achieved by trade alone, unless new goods can be brought into the economy from somewhere. New goods are produced by technological means – including agriculture and forestry – either in the domestic economy or are brought in by traders from foreign lands.

The economy of a country provides and distributes a great variety of goods, largely produced by technology. Of no less importance, however, is the fact that the economy provides employment. The importance of employment is twofold: it provides income for the employed, or self-employed, person and it provides the person with a place in society and a sense of being useful. An unemployed person is deprived of a large part of his/her income and is deprived of the company of fellow workers. The companionship of workmates is of great importance to most workers and the private social life of many people is enhanced by, sometimes dependent on, contact with fellow workers. The unemployed find it hard to structure their day. Why should I get out of bed and how should I fill all these long vacuous hours before I can go back to sleep? Last, but not least, the unemployed lose their sense of self-worth. Nobody needs me, I contribute nothing to the world, I am useless and worthless.

Work and employment are the most valuable things that technology both provides and takes away. We should do all we can to maintain employment as one of the most valuable goods. In the relentless pursuit of profits managers should do all they can to keep their workforce in employment and use redundancy only as a very last resort. Unfortunately managers often do not see things quite this way.

---

5 We distinguish between absolute and relative poverty. Somebody absolutely poor suffers real deprivation and may suffer from hunger and cold; somebody relatively poor simply cannot afford all the goods and services that most people enjoy.

Over many years, and particularly between 1950 and the present day, employment patterns have changed dramatically[6]. Whereas in 1900 agriculture provided about 11% of UK employment, the share of agriculture in employment had fallen to about 2% by 2000. During the same period the share of employment in manufacturing industry fell from 28% to 14%. Employment in services increased from 34% in 1901 to 75% in 2000.

The causes for these dramatic changes are not hard to find. Undoubtedly the main cause was technological change. Technology changes as a result of both internal forces, such as curiosity research, ambition, and, first and foremost, the constant thriving for profit. Some technological change is caused by interactions between technology and society. Thus social forces can also be causal factors in technological change. The trend to the nuclear family and the trend toward equal rights for women are closely related to the introduction of modern household machinery, as is the trend away from employment of domestic servants. Technology in these situations is not the cause of change, but plays the role of an enabling factor. How the relationships work in detail will not here be investigated. In oversimplified terms we may say that domestic machinery has freed women from much domestic drudgery and has encouraged middle-class households to manage with only casual part-time hired labour. The relationship between the trend toward the nuclear family and technology is complex and beyond the scope of this book. Emancipation of women and enhanced mobility no doubt play key roles.

The technological changes that brought about diminishing employment in agriculture and, to a lesser extent, in manufacturing industry are, in a nutshell, mechanisation, followed by automation. When sowing by hand was replaced by machinery and the horse-drawn plough gave way to the plough pulled by tractor, and when harvesting was no longer carried out with the scythe followed by laborious threshing, and when all these operations were replaced by combine harvesters requiring only a fraction of the labour used previously, it is not surprising that the total labour force required by agriculture shrank to a fraction of the previous force. With a workforce in agriculture, forestry and fishing of only 2% of the total UK workforce, the contribution of these activities to GDP is also about 2% of total GDP.

A further factor that contributed to greater efficiency in agriculture, meaning the creation of greater value with less labour input, is plant and animal breeding in the widest sense. Thus modern farm animals produce far more of whatever they are supposed to produce with the same inputs of labour and materials. Modern plant crops produce greater yields per unit area. Tree felling and the preparation of timber products has been largely mechanised and thus less labour is needed for the same output. Fishing has also increased its productivity with the calamitous result of over-fishing and the threat of extinction for many species of fish.

Work in agriculture has fundamentally changed. Gone are the days of extremely hard labour with pick and shovel and scythe and horse-drawn plough. The farmer or farm labourer now works with sophisticated machines, the cow-shed is often automated, and the computer plays a central role in farm management. Brawn has given way to brain.

The changes in manufacturing industry are every bit as dramatic as those in agriculture. Not only has total output of British manufacturing industry, particularly of British-owned manufacturing industry, declined by a huge amount in favour of foreign-owned industry and of imports, but industrial productivity has also increased owing to mechanisation and automation. Some branches of industry have declined dramatically. Coal mining has almost disappeared, ship building has declined, general engineering has declined, the British-owned motor industry has disappeared and total output of British industry is much diminished, with the gap being filled by imports.

Gone are the days of factories employing thousands of workers on a single site and gone are the days of whole villages and regions living on coal-mining[7]. In the fifties and sixties a production line in an automobile factory employed hundreds of workers each placing a part onto the car being produced and screwing, riveting or welding it on by hand. It looked very much as parodied in Charlie Chaplin's *Modern Times,* whereas a modern automobile production line is operated largely by robots and other automated machinery. Work is almost

[6] See e.g. March 2003, Craig Lindsay, A century of labour market change:1900 to 2000, Labour Market Trends, pp. 133–144

[7] Between 1917 and 1927 over a million coal miners were employed in British collieries. In 1947 the National Coal Board took over the management of 958 collieries employing 703,900 miners. In 2001/02 only fifteen deep mines remained employing just over 6000 miners. In addition there were 20 opencast mines in Britain.

as monotonous as before, but physically less demanding and much more effective. Similarly, coal is no longer hewn out of a coal-face by pick-axe or hand-held pneumatic drill and then shovelled onto a conveyor belt to a railroad, where it was loaded by hand onto wagons and transported to the lifting gear. A modern mine operates with huge automated machinery that enables a handful of workers to produce vast amounts of coal. The work is far less strenuous and less dangerous than in the past and, of course, much more efficient in terms of input of labour. On the other hand, narrow coal seams cannot be mined by modern methods and production has to be concentrated on relatively few favourable spots. The trade-off is that mining is no longer a hazardous and back-breaking occupation, which surely is a very good thing. The major costs in mining are no longer labour costs; capital costs are now dominant.

Coal has acquired a bad name as an environmental pollutant. There is no doubt that burning fossil fuels, such as coal, inevitably produces carbon dioxide, the gas that is regarded as the main cause of global warming. Concern about the increasing use of coal in China and India is widespread. However, modern research has developed methods whereby the burning of coal on a large scale, as for example in electricity generation, need not be environmentally harmful. It is possible to collect the carbon dioxide produced in the combustion process and bury it deep underground or in disused oil and gas fields in the ocean-bed. The method is known as carbon capture and storage (CCS) and Norwegian experience has shown it to be practicable and effective.[8] These methods are not fully developed yet and are still expensive, but they look very promising and might enable many countries to use their ample coal reserves for the generation of electricity without causing harm to the environment. The highly problematic and controversial nuclear option is by no means the only option for generating electricity without causing global warming. There are all the possibilities of using renewable energies, but there is also a possibility of building and operating clean coal-based electricity generating plant. Opponents of coal argue that such plants would be expensive. In view of the fact that the cost of nuclear power stations is somewhat indeterminate and almost certainly much higher than its proponents admit, coal is very likely to be rather competitive and using coal might be a sensible use of rich domestic resources in many countries, including Great Britain and the US.

The shrinking agricultural and industrial sectors made room for the expansion of public and private services. Employment in the service sector has risen to 75% of all employment and the contribution of services to GDP has risen to 66%. Public services have expanded considerably, despite the new-found religion of privatization. Social services and the National Health Service have become major employers, and education has expanded markedly. In total, government expenditure as a proportion of GDP has gone up to around 40%.

There is nothing wrong in principle with an expanded service sector. However, some services are more useful than others. The British service sector became dominated by financial services that appeared to make a large contribution to GDP, and while there was an abundance of financial advisers, it was hard to get a plumber. Worse still, the financial services sector appears to have operated a series of scams and its huge profits were made by conjuring tricks producing money out of thin air. Now that the sector has virtually collapsed and its fat cats are licking their wounds[9] instead of carting off their unjust rewards by the lorry-load, the real economy is being dragged down with the collapsing financial institutions and governments have no option but to produce rescue packages financed by issuing new money or government bonds (gilts), more or less eagerly bought by the wary public at home and abroad. When the going gets tough, the majority of even the most enthusiastic free marketers remember the theories of John Maynard Keynes, calling for government expenditure in times of a market slump and begin to appreciate the need to set bounds to the absolute freedom of markets. The call for more and better regulation comes from many quarters and it is to be hoped that the wild excesses of the banks will be curbed and some honesty and sense will be brought back into the financial sector.

It is often argued that services are as good as manufacturing to sustain an economy. This may be true for a small country in a special situation, such as having a thriving tourist industry or exceptionally profitable financial services. Countries such as Liechtenstein, the Cayman Islands and the British Channel Islands come to mind as examples. During the Thatcher years it was argued that the City of London is the greatest British asset

---

8 See e.g. The Observer, 15. February 2009, pp. 18–19

9 It would appear that some of the fat cats are managing to embark upon one of their legendary spare lives

and it was thought that letting British industry decline does not matter. There were always those who regarded these policies as misguided, but just how very badly misguided they were, is now becoming crystal clear.

An economy relying in large part on honest services, such as medical services, education, social services, catering, transport, and craft services such as plumbing, joinery or hairdressing, alongside some industry, could probably work reasonably well. An economy relying heavily on financial services, to the exclusion of almost all else, blatantly cannot and this is the reason why Britain is being hit so very hard by the current world financial crisis. The banks could benefit the economy at large if they fulfilled their basic function of providing credit where credit was needed and if loans were made with due regard to the ability of the borrowers to pay interest and repay the loans; in other words, if banks acted with proper diligence and honesty. When banks and other financial institutions marketed loans quite recklessly, pushing hapless people into purchasing everything from houses to cars and consumer goods on the never-never, a huge bubble of debt grew ever larger until it reached bursting point. What is now known as sub-prime lending was in fact reckless lending that was bound to end in disaster. Disaster did indeed strike in due course, in late 2008. At the time of writing, in early 2009, the disaster is in full swing. The total credit market debt as a percentage of the total economy (GDP) in the USA stood at about 160 in the year 1980; by the year 2006 this ratio had increased to about 330.[10] This, it would seem, is far too high for a healthy economy and eventually many citizens were forced into default and, in the worst cases, lost their homes.

It now turns out that banks can not only make large profits, but can also accumulate losses of magnitudes that exceed by far the imagination of most people. Making a loss in business is always regrettable, but not necessarily reprehensible; making astronomical losses from reckless lending to private individuals and businesses is more than reprehensible. In the eyes of the law-abiding citizen it is regarded as a result of criminal activity, even if no laws were actually broken. Unfortunately *due care* is not a law that bankers appear to be bound by. The banks have caused untold misery to a lot of people and are now being bailed out by governments that can ill afford to raise such huge sums of money, especially as some of the promises to insure banks against losses are equivalent to throwing money down a bottomless pit. It seems likely that some of the finance now raised by governments will never be paid back, as governments are entitled simply to create money via their central banks and pump it into the economy. In sum total the present operations of governments, necessary as they are, will cause considerable future budgetary problems.

The behaviour of investment bankers has in no small measure contributed to the mistrust of ordinary people toward those in authority in general and bankers in particular. It is unclear whether the behaviour of banks is just a facet of unfettered capitalism that has caused a lack of trust toward business, authorities, politicians and everything else, or whether banks act in a powerful world of their own. I think that a free market economy can only function properly if it adheres to strict rules; a totally free market, especially a totally free global financial market, leads to excesses that cause first a loss of faith and eventually lead to total calamity.

Central banks are different from all other banks, as they are supposed to create money, act as lenders of last resort by lending to banks at the interest rate that they fix from time to time, and try to control inflation. In recent years and months, the control of retail prices in the major economies was reasonably successful. Indeed free markets in consumer goods are frequently reasonably stable of their own accord. Markets for labour, land, housing and capital are, on the other hand, inherently unstable. Unfortunately this is not accepted by classical economic theory and central banks have not attempted to stabilise prices for such goods[11]. As a result of such lack of attention, there has been a veritable explosion of house prices and of share prices on the stock exchange. A majority of the money in circulation in such markets comes from credit and thus the credit in most highly developed economies has expanded beyond reasonable limits. Indeed a credit bubble formed and this bubble has now burst. Much credit was extended without due care and the stock exchange prices, also driven up by credit and by complex machinations, burst together with the credit bubble.

The old-fashioned banker was mostly honest and modest, content to safeguard the funds deposited by savers and making loans to customers who could, by and large, be trusted to pay interest and repay the loan. From

---

[10] Data from Ned Davis, Inc., 2007

[11] 2008, George Cooper, The Origin of Financial Crises, Harriman House, Petersfield, UK.

the earliest days of banking, when gold was the common currency, bankers lent out more than the deposits they were holding. Experience taught them that they needed to hold only a fraction of total loans as a reserve for those customers who wished to withdraw their deposits. Thus a risk of a run on the bank was always present, but the degree of allowable risk was based on experience and no elaborate risk assessment calculations were required. The replacement of gold by paper money made no difference; so-called fractional lending remained the norm. It is a large leap from the relatively simple banking to modern banking and financial institutions. We now have a multitude of financial institutions covering a multitude of financial services. Some might say that we now satisfy a multitude of financial needs, but I regard many of these so-called needs as commercial, frequently unsound, offerings rather than true needs.

The major financial institutions are the stock exchange, commercial banks, investment banks, insurance companies and specialist mortgage lenders. In reality many institutions combine several of these activities. We may regard all financial services and transactions as information services and should not be surprised that modern information and communication technologies have become indispensible, and indeed dominant, in the financial sector of modern economies. The introduction of the computer and of computer experts has fundamentally changed financial activities. Bankers became bolder in their taking of risks because they thought that with the aid of computers and sophisticated mathematical models they could fully understand and control the risks they were taking, and remained convinced that those who dare will win. In other words, bankers began taking great controlled risks, producing large profits. In reality you cannot control risks; at best you can calculate the probabilities of certain events happening. Investment bankers became highly inventive in producing new risky and profitable financial instruments, far removed from the original simple practices of bankers. The mathematical theories on which some novel financial instruments are based became highly complex and demanded highly skilled computer experts. The financial transactions enacted by investment banks became very close to those of a casino. Investors were not sold shares in real tangible businesses; they were sold odds on future prices of commodities or shares. Thus the financial sector became almost totally divorced from the real economy. Although shares dealt on the stock exchange still represent a slice of ownership of real businesses, the values assigned to shares are based on expectations, and have little or nothing to do with the real worth of these businesses. Worse still, the new financial instruments, so-called derivatives, represent nothing real at all; they are sophisticated bets based on sophisticated computations.

Another important effect of modern technologies on the financial sector is the interdependence between institutions all over the globe. Banks and other financial institutions are inextricably linked by a web of mutual investments, credits and management. Because the use of modern computer and communications technologies (Information Technologies) makes global transactions practically instantaneous, the global financial networks have become virtual organisms with huge financial flows connecting the different parts. In the ten years between 1995 and 2005 gross flows of financial capital between countries have almost trebled to reach 16% of world GDP.[12] Because of these linkages, the failure of one large institution is likely to cause failures in further institutions and the whole financial system becomes unstable.

As the current credit crisis has shown, the problems of the financial sector very quickly affect the real economy. Though so many financial transactions are unrelated to the real economy, nevertheless the real economy is vitally affected by problems in the financial sector. For business of almost any kind to thrive, it needs frequent favourable credits from the banks; for the housing market to thrive, it needs reasonably priced and freely available mortgages. For the consumer sector to thrive, consumers need credit for large purchases. Unhappily in the recent past prices of houses rose dramatically and mortgages were given to people unable to meet payments of interest and repayments of capital. Equally unfortunately, consumers bought too much on credit. To compound the difficulties many financial transactions, such as the purchase of shares on the stock market, were also transacted on credit. Thus a bubble of credit grew and eventually burst, when too many borrowers of all kinds became unable to meet their obligations.

In recent months a number of investment schemes, some of them of very large dimensions, turned out to have been fraudulent. Many bona fide investors have lost their money and these fraudulent operations have not

---

12 www.theage.com.au/news/business/globalmoney flow soars.....28.01.2009

exactly enhanced the reputation of financial institutions. Yet again it has been shown that excessive greed and "wonderful opportunities" have presented bankers and their likes with "irresistible temptations". It is clear that inherent honesty and decency is in inadequate supply in financial circles and that it is vitally necessary to supervise and regulate their activities much more closely and carefully. Where honesty and morality fail, the law has to step in.

Bankers have caused a substantial loss of trust and this deficient trust is a very serious deficit indeed. Society cannot function properly without an adequate measure of trust. We need to trust our institutions, and, at a more basic level, we need to trust all people we interact with, whether in commercial or social intercourse. Unfortunately practically no modern society is achieving a perfect degree of trust because all societies contain criminals of various kinds. But the present loss of trust in most of our institutions, and in many leading figures in our society, affects us all and isolates and alienates us from our society. To be fair, it is not only the fault of the bankers. It is also the excessive speed of change, including technological and commercial change. How can we trust products that change all the time and that we cannot adequately test before buying them? How can we trust firms that are in continual flux by being taken over or merging with other firms? How can we trust institutions that are continually reformed and never settle into a proper working mode? Perhaps President Obama was right when he emphasized the need for change in the USA, but we can have too much change and we have a great need for continuity and for trust in existing institutions and procedures.[13]

In some services it is difficult, if not impossible, to replace human workers by the deployment of technical means. Nursing, teaching, care for the young or the elderly are examples of services that are inevitably labour intensive. Because productivity in such services is low, they are expensive to run, difficult to staff and difficult to finance. This is highly problematic, because such services embody the very essence of being humane and civilised. The care of the old and the young, education and health are the Cinderella of capitalist technological societies. Care for the natural environment is another of the forgotten children of such societies, although it is possible to make profits from the introduction of new technologies that combat environmental degradation. The creation and introduction of environmentally friendly technologies could produce very many opportunities for profitable investment and for employment. Out of the many possibilities, suffice it to mention improved insulation for houses, more efficient motor cars, better public transport, better methods for recycling and cleaner electricity generation.

Change is, and always has been, an essential characteristic of human society. Changes in technology – technological innovation – differ in some respects from general social change in that it is more unidirectional and more rapid. Technological innovation is deliberate and, because of the prevalent faith in its economic benefits, it is regarded as highly desirable. The desirability of innovation seems to have been transferred from the sphere of technology to most other spheres of life. It so happens that technology, by implication, makes all novelty desirable. We may speculate whether the acute search for novelty in technology has influenced the general desire of contemporary society to seek new ways in every endeavour. Indeed novelty has largely replaced criteria of quality; if it is novel it must be good. This quest for novelty often leads to absurdities, especially in the arts and in the media, but also in education and other fields.

Video techniques allow all kinds of tricks, such as overlapping semi-transparent pictures, and rapid sequences of images that confuse the viewer and make no sense. Opera and drama have been infected by a relentless search for novelty of staging that often flies in the face of a sensible and sensitive interpretation of the intentions of the original playwrights or composers.

It may be that technology is to blame for some of the new-fangled ideas because using new technology makes it possible to tread entirely new paths. But quite apart from the fact that technology has made some things possible that previously were not feasible, technology seems to have infected all our thinking with the desire for novelty even in spheres where quality, rather than novelty, should be the measure of things. No doubt the application of new technologies opens up new possibilities that should be used, but they should be used with the guidance of criteria other than novelty. Not everything that can be done should be done. Whereas technological innovation, with all its aberrations, makes at least some economic sense, the general elevation of

[13] The dangers of rapid technological change are discussed in 1995, Ernest Braun, *Futile Progress*, Earthscan, London

innovation per se as an attribute of value makes, in my view, no sense whatsoever. The arts and, to some extent politics, are particularly affected by a pathological desire for novelty, forgetting all else in the quest for originality. Alas, novelty cannot be a substitute for quality!

One of the most fundamental laws of technological progress states that all technological development is unidirectional in the sense that all new devices, replacing older technology, show improvements in performance. They may be more efficient, more convenient to use, more accurate, perform tasks that the preceding technology was unable to perform, or be "better" in some other way. More recent developments of technology cast doubt on the universal validity of this law. We find some changes in the design of various consumer technologies of very questionable value. Unfortunately the law does not hold absolutely, some designers indulge in useless gimmicks to induce purchasers to buy their goods. Take, for example, recent developments in small digital cameras. They no longer provide viewfinders and use a screen instead. As a result, it is impossible to use them in strong light – no more photographs of kids in the sunshine. They are less convenient to use than older models and this lack of convenience is not balanced by any other improvement. This development is the result of an attempt to create a new fashion in cameras, and by creating a new fashion, making older models obsolete in order to create new markets. The new cameras are as unreliable and as short-lived as their immediate predecessors. Perhaps manufacturers deliberately make small digital cameras difficult to use in order to enhance sales of more expensive digital single lens reflex cameras.

There is a general assumption, and frequent assertion, that all technology is created to satisfy some human need. We can, of course, define need in such ways as to make this dictum true, but that would lead to logical absurdities and would turn the relationship between human need and technology on its head. The example of a recent technology that most definitely was not developed in response to any reasonably defined need is the introduction of very small cheap motorcycles, unsuitable as road transport and intended to make money out of selling these lethal machines to teenagers or children. They are quite fast and unsafe and can be bought by children from their pocket money or by indulgent and ill-advised parents. The "need" they satisfy is to have fun through speed, noise, and danger and the "need" to impress the peer-group through misunderstood manliness. These machines cause injuries, and even fatalities, and are a nuisance to all those unfortunate enough to be near the scene of their activities. To my mind, it is an irresponsible development and fulfils a need only if we pervert the meaning of the word and designate everything people buy as a need. This is precisely what uncritical supporters of free market economics do. It must be admitted that as far as economists are concerned there is no other option than to regard all purchases as needs because neither statistics nor economic theories are able to distinguish between necessary and foolish purchases. Economists have no option but to regard all purchases by consumers as satisfying needs, from the macroeconomic viewpoint it is not possible to differentiate between purchases according to the motivation of the buyer.

Increases in productivity, i.e. more products produced with the same input of labour and other factors of production, can also lead to economic growth, provided the savings in labour lead to increased production rather than to unemployment. Unemployment is not only a great waste of human resources and a drain on state finances[14], it is also a personal tragedy for those willing to work and unable to find employment or create income from self-employment or other legal activities. The unemployed lose their main link to society, become socially isolated and find it hard to structure their time in any satisfactory way. They suffer all this in addition to suffering financial hardship.

There is one special case in the provision of goods and services: the provision by the state. In most countries the state, or its agents, provide essential infrastructure, such as roads, sewage, street lighting and cleansing and, until very recently, electricity, gas, telecommunications, postal services, railways, ports, and airports. In many countries the ideology of privatising as many as possible of these infrastructural services has become dominant and they are now often provided by private enterprise under some form of state supervision. Thus the providers of many essential services now use technology in the pursuit of profits, rather than for the exclusive purpose of providing essential public services or public infrastructure.

[14] In those states where unemployment benefit is a right paid by the state

A case in which the division between private and public provision has long been problematic and controversial is the provision of health services. In recent times medical technology has made very large strides and this has caused the cost of providing up-to-date medical services to rise very rapidly. There is a lot of money to be made out of innovative medical technology and, as a result, the provision of full medical services for whole populations, without discrimination between rich and poor, has become problematic. Of course not only medical technology has advanced, but the pharmaceutical industry has also not stood still. There are many new drugs on the market and many of them are extremely expensive because, so the industry claims, research into new drugs has become complex and costly.

Finally, we must mention the armed services, including both internal and external security services. Virtually all states provide these services out of the public purse, though private enterprise provides some supplementary services. Innovation in arms and armaments and other military technologies is very rapid and the cost of armed services is high. Industry that supplies the armed services often makes very large profits, thus technology is here again seen to be in pursuit of profits. Yet it needs to be said that governments put pressure on industry to produce improved weapons and weapons systems, as superior weapons can provide a decisive advantage in modern warfare. The arms race favours the richer nations and often beggars the poorer ones, and the arms industry and arms dealers make profits out of all clients, rich and poor. As we live in a world perverted and primitive enough to settle its disputes and pursue its political aims by the use of arms, we may say that the need for arms is a real social need. I regard this as extremely unfortunate and regrettable, but cannot deny it. The use of the most sophisticated weapons in the pursuit of the most primitive of human activities – warfare – is the most disgraceful absurdity of our time.

The tradition of warfare is at least as old as the oldest civilizations, and possibly older. The early civilizations began a tradition that is still alive: the tradition of warfare with the purpose of increasing wealth and power of the aggressor at the expense of other people. All ancient states kept armies and used them to attack neighbours near and far with the aim of plunder, of acquiring slaves, of exacting tribute, of expanding their territory and of self aggrandisement. We speak of great empires and great rulers, more often than not meaning their military might, their cruelty, their wealth, and their successful military campaigns. The word "great" is here used in a perverted Pickwickian sense. Unfortunately the connection between technology and warfare has always been, and still is, extremely close; they are, indeed, inseparable twins.

Technology in all its aspects has become an essential economic activity. We spend a lot of effort on introducing new and improved manufacturing technology in order to streamline the production of goods, we constantly attempt to introduce new or improved products in order to stimulate flagging markets, and we constantly strive to improve technologies used in services and in administration in order to improve some services and streamline administration. As technology has become a central economic activity, it should not surprise us that the measure of all economic activity – money – is also the measure of all activities relating to technology.

It is true, of course, that the vast numbers of people, and the complex societies they live in, could not be sustained without the extensive use of technology. But it is equally true that most contemporary technological innovations aim primarily at making money rather than at improving our lot.

All producers of technology depend on selling their goods, in other words on customer demand in the strictly economic sense. Demand is what people buy. The causes of demand, i.e. the reason why people buy certain goods or services, may be ordered into several categories. We have mentioned purchases by private individuals and divided these into purchases by different social groups. We now turn to purchases by organisations.

1. *Purchases by social organisations.* Government in all its forms is a major buyer of many goods and services. In recent times many government functions have been privatised, but in these cases private firms simply fulfil roles previously fulfilled by Government. The trend continues and some writers call the phenomenon of a government that has passed many of its functions to private enterprise a 'hollow government[15]. From the point of view of the Neo-Conservatives the hollow state is highly desirable, to less radical right-wingers and, of course, to left-wingers, such a state is a horror vision. Government buys directly or by proxy of private firms major and minor equipment for its armed forces, for its transport networks, for the police, for

[15] See Naomi Klein, (2007), The Shock Doctrine, p. 294

public administration and so forth. No matter who the purchaser is, a functioning state needs large inputs of goods, many of the high technology products, and many services. Indeed public administration may be viewed as a service. Though many of its functions are performed by civil servants, quite a few functions are bought in, even in a state in which privatization has not been driven ad absurdum.

2. *Purchases by industry and business.* A very large amount of buying and selling, of supplying and consuming, takes place between commercial organisations. If we take the car industry as an example, we find that the motor manufacturer mostly carries out the assembly of parts supplied by a whole host of suppliers. The car manufacturer coordinates all the functions necessary for the production of cars, buys in the parts and often even the designs, assembles the cars and sells them to the public. Thus the manufacturer may be regarded as a large scale purchaser of products and services that he assembles and markets, where marketing is in itself a major function.

One of the outstanding characteristics of technology is that it consists mostly of systems[16]. There are two kinds of technological systems. On the one hand, we may speak of, say, a system of domestic technologies consisting of washing machines, dishwashers, cookers, vacuum cleaners and so forth. Each of these machines is relatively self-contained, though all depend on external inputs. The dishwasher, for example, requires various ancillary inputs such as running water and an electricity supply, salt, rinsing aid, and special detergents. It also requires adequate drainage, and we may call this assembly of the dishwasher with its inputs and outputs a sub-system of the household machinery system. When we speak of a system of household machinery, we mean an assembly of machines and devices that help us run a household; yet we do not mean a system of interdependent technologies. We may run a washing machine on its own, without necessarily deploying any other household machinery, but not without some external inputs, such as detergent. As an aside we note that most household machinery merely performs tasks that we can perform without machinery, merely by the labour of our hands, aided by simple tools. We can wash dishes and clothes without machinery, we can brush and beat carpets, we can cook on an open fire. Household machinery is typical of the class of technologies that make life easier and save labour, yet has no capabilities entirely without the range of human abilities.

The concept of even such loose systems – more assemblies of machinery rather than systems – is helpful in the search for spotting marketing opportunities for new technologies. The potential innovator surveys the "system household" and seeks gaps in the provision of technical aids, as well as improvements in existing machinery and gadgets. The search for new opportunities is facilitated by the system nature of technology, even in cases where the system is only a loosely interconnected one. Each generation of domestic machinery is a little better than its predecessors and gaps in the mechanisation of the household are quickly filled. More recently a whole new system of domestic communications, electronic data processing and entertainment has been added to the common range of household equipment and more is being added all the time. The complete system of household technology consists of domestic machinery, household electronics, and furniture and furnishings, not to mention the dwelling itself with its sub-systems such as heating, water supplies, electricity supplies, roofs, insulation and so on.

Many technological systems are much more close-knit. The number of examples is well nigh infinite. We shall describe one or two examples, but only after describing such systems in general. A technological system consists of one or several interdependent devices that require for their production and their use a number of additional technological inputs and the fulfilment of some technical and social preconditions. Take as an example the ubiquitous computer. In practical terms, the computer could not be developed without some mathematical analysis and without the availability of semiconductor electronic devices[17]. These devices were initially developed as replacements for cumbersome electronic elements known as tubes or valves, without initially thinking of computers as their eventual main field of application. Indeed the computer could not have become the universally applied tool without the prior development of integrated circuits, making it possible to gradu-

[16] There are many definitions and many kinds of systems. For our purpose a system is a functionally related group of elements that perform some specified function.

[17] For a description of the initial history of solid state electronic devices, so-called chips, see e.g. 1982, E. Braun & S. Macdonald, *Revolution in Miniature*, 2nd ed., Cambridge University Press, Cambridge

ally increase the computing power and computing speed by giving individual chips enormous computing, switching and memory capacities. To do all this, a large array of technological developments were necessary. The development is not, of course, finished and both individual chips and computer systems are still becoming increasingly powerful.

We cannot possibly enumerate all the items and sub-systems that go into the making of a computer and shall mention only a few essential ones. To make modern computer chips it was necessary to develop the art of purifying materials, especially silicon, and the art of growing large near-perfect single crystals of this material. To manufacture chips from the crystals they have to be sliced and so-called doping materials have to be incorporated in very precise patterns on a very small scale. For this purpose masking, etching and diffusion or ion bombardment techniques were developed and, of equal importance, highly sophisticated masks had to be produced photographically to define the areas into which diffusion should take place.

Even the most sophisticated computer chips would be useless unless the art of programming computers had been developed to a very high degree of sophistication and efficiency. A computer has to be able to store its programmes, as well as a large amount of external and internal data. So we need different memory devices, and we also need devices to allow the input and output of data. Finally, we need to be able to print or transmit the outputs from the computer. This requires the development of visual output devices, such as cathode ray tubes or liquid crystal arrays, and the availability of fast printers.

This very sketchy and incomplete enumeration of the various technical sub-systems that go into a computer system should suffice to illustrate the essential nature of computer technology as a technological system, consisting of much varied hardware and a great deal of software.

Complex as all this is, it is not enough to make computers operational on the vast scale on which they currently operate. To make this happen, some social conditions had to be fulfilled. The principal social ingredient that had to be provided was a trained workforce able to produce, operate, program, service, repair and sell computers. The very first computers were designed by mathematicians and the first chips were produced primarily by physicists and chemists. The pioneers in every new field have to acquire their knowledge by trial and error and are essentially self-taught. These early experts then train the many new experts required, until a whole system of training and qualifications is established and a proper job-market develops.

The most crucial and essential condition for the establishment of a new technological system is the willingness of entrepreneurs to invest money in new ventures. Production facilities have to be built and a market in the new devices needs to become established. Both the risks and the potential gains are very large while the first uncertain steps on the path of a new technology are taken. In the case of semiconductors and computers the public purse helped to take some of the risk out of the enterprise. The military eagerly bought the new devices at prices that few civilians were prepared to pay. At a somewhat later stage, governments provided all kinds of help to the emerging technology because they became convinced that the economic future of their respective countries was at risk unless their country mastered and established the new technology at an early stage.

Finally, the computer had to find acceptance by a large number of people in a large number of applications. The technology had to become easy enough to use by only moderately trained people and its advantages had to be large enough to outweigh its disadvantages. This is a universal rule for the diffusion of new technology. People are always reluctant to change their accustomed ways and to throw overboard their skills and their experience. New technology has to offer considerable advantages to overcome the natural reluctance of people to change their ways; or so one would think. The most recent social climate, the'Zeitgeist', helps to overcome these natural human and organisational tendencies. The current Zeitgeist makes much of the value of progress and of modernity, whereas being old-fashioned has almost become a term of abuse. To be modern, progressive, 'with it' are desirable attributes, whereas being conservative, unadventurous, and reluctant to accept change are regarded as wholly undesirable attitudes. The ode on progress is sung by all and sundry, is heard from every rooftop and, of course, from all the political, cultural and – last but certainly not least – commercial establishments and the media. Old is out, novelty is king. Progress is the supreme goal. And, because technical progress is so much easier to achieve and to measure compared to social progress, technical progress has become virtually synonymous with progress as such. I regard this as unfortunate, because we could well do with social progress. Technical sophistication is no substitute for what has been called "the good society".

Taking technology as a whole, it is correct to say that it consists of an assembly of technological systems of various sizes, various degrees of complexity, various degrees of interdependence and various degrees of utility. All that has been said about household technology or computer technology as examples of technological systems could be said with equal validity for systems such as railways, road transport, aircraft, machine tools, textiles and clothing, and many more. Modern technology encompasses so many systems and sub-systems that any enumeration would be tedious and boring; suffice it to say that the totality of technology may be divided into a large number of systems, sub-systems and, within these, individual technologies. Each of these systems serves a particular group of purposes. These can vary from extending human capabilities beyond their natural limits (e. g. flying), or simply to ease the burden of physical labour and/or to accelerate the performance of tasks that are within human capabilities, albeit aided by simple hand-tools or implements, (e.g. digging, washing clothes).

Many complex novel technologies are troubled by unreliability. Actually it is the consumer who is troubled by malfunctions and breakdowns of many apparently wonderful gadgets. And, because there is an enormous discrepancy between the productivity of factory systems of production and repair workshops, repairs to malfunctioning gadget are comparatively very expensive and often hardly worthwhile. Quite a few relatively new but malfunctioning products are thrown away rather than repaired – we have become a throwaway society in many respects. This leads to waste of materials and to a good deal of frustration.

Some technological innovations are so radical that they form the beginning of an entirely new technological system. It is difficult, often impossible, to foresee the development of a radically new system when the radical innovation is still in its infancy. Examples are easy to find. The invention and introduction of the automobile started a radically different system of road transport and a very large new industry. In its further consequences the automobile caused fundamental changes in the way we live. If we consider the technological system "automobile", we need to consider the manufacture of thousands of individual parts that go into the assembly of a motorcar and the process of planning, designing, assembling and selling motor vehicles. We further need to consider the system of roads and all their ancillary features, such as the production of tarmac, traffic lights, bridges, tunnels, and so forth. And, of course, the system of producing and delivering fuel. The introduction of motor vehicles has brought in its wake a whole range of new legislation and regulation, including driving licences, compulsory motor insurance, vehicle inspections and vehicle licences. The automobile revolution has changed the face of towns and we may claim, without fear of exaggeration, that it has changed the face of the earth as well as the structure of society.

Major social change follows in the wake of major new technological systems. We have discussed briefly how the motorcar has changed the life of individuals, has contributed to change in family structures, in retail shopping, in town planning, in patterns of consumption and in environmental hazards. We should add a brief reminder that the widespread use of the automobile has also influenced global politics. Oil has gained fundamental importance in our economies and the politics of oil exploration and marketing has become a major factor in international relations and politics. Although no politician has admitted it, there is a widespread belief that the recent war in Iraq was fought primarily in order to secure Iraqi oil supplies for the USA and Western oil companies.

The most radical single innovation that contributed substantially to the rise of the computer to a major technological system was the humble transistor invented in 1948 as the forerunner of semiconductor electronics. Semiconductor electronics became a major industry in its own right and is at the heart of all modern computers.

The widespread use of computers has caused manifold changes in society. We have mentioned that the combination of data processing and communications technologies has enabled huge multinational corporations and financial institutions to be effectively controlled. This, in conjunction with modern transport technologies, has led to the phenomenon of globalisation, i.e. to the dominance of relatively small numbers of international corporations over most of our economic activities.

It is virtually impossible to know from the outset that a particular technology will lead to a huge system with enormous social repercussions. At the beginning of a radical new technology the scale of things to come is always underestimated. When the first automobiles came on the market, it was thought that they would serve only as toys for the rich. Nobody envisaged the automobile at first as a means of mass transport and nobody imagined that the automobile industry would grow into one of the largest and most important industries in the developed world. The same happened with the first computers, with the telephone, with solid-state electronics, and so forth.

The radical innovation that becomes the forebear of a new technological system must have the potential of substantial utility and of capturing large markets. It will either replace older technologies (e.g. the automobile replacing the cart and horse), or fulfil some task that no other technology had fulfilled before (e. g. the aeroplane or the computer).

The new technological system may arise out of a single invention, but generally it will use older technologies in addition to the new invention for its early implementation. In the further course of its development the radical new technology will draw on numerous further new inventions for its eventual development into a new system. As soon as the potential of the new system becomes apparent to investors, there will be a rush of new investment that, generally speaking, will overshoot the mark. Once the potential of a new technology is realised, many old and newly founded firms endeavour to produce the new technology and try to reap above average profits from the novelty. Sometimes this pays off, but generally speaking many of those who chase after the novelty fall by the wayside and may not even recoup their investment.

The birth of a new technological system conjures up the image of a river. Even the largest river starts from small sources. It grows into a brook and on its way absorbs tributaries and thus swells into a river. Eventually it may become a major waterway with profound effects on the landscape and on the population that lives near it.

The large-scale application of a new technological system is invariably associated with radical social changes. Apart from the usual changes in occupations, it may affect the geographic distribution of populations, trade flows, and the very essence of personal and family life-styles. If we compare human society as it was roughly a century ago with human society today, we observe huge differences and note that many of them were caused, directly or indirectly, by road transport in general and the automobile in particular. Similarly, telecommunications had a profound effect upon society, as did air transport, television and the ubiquitous computer.

One of the properties of technological systems is that they can be both perfected and expanded by additional innovations. Once a new system starts on its course, numerous potential innovators look at every aspect of it in search of new opportunities to make profits. These may arise out of possible improvements in the manufacturing process of the new technology or of improvements in its performance. They may also arise by expanding the system to perform tasks that were not originally foreseen or envisaged.

Generally speaking, technologists and industrialists are always on the lookout for innovations that they might persuade the public to buy. The existence of a relatively new technological system that can be both improved and expanded is a great help to their endeavour as it focuses their search for innovative potential. Indeed occasionally an innovation is positively demanded by industry and we speak, rightly, of demand-driven innovations. This is the case when a new device or system shows a definite weakness, if there is an identified missing or weak link in the system. In early computers, even when solid-state electronics had overcome the greatest weakness of unreliability and large power consumption of thermionic valves, there were weaknesses in memory devices and in printers. On the other hand, when it comes to expanding the capabilities of technological systems, the innovations are more likely to be of the technology-push type, though with relatively easily foreseeable market chances. When large expensive computers had established themselves, it did not require great foresight, nor did it involve great risks, to attempt to market smaller and cheaper computers. Similarly, it is fairly clear that increasing the capacity of memories and increasing the speed of computers would be accepted by a market always keen on cheaper models and improved performance of virtually all technological devices.

The humble telephone provides some good examples of the ways in which a nascent technology can be gradually improved and expanded. The first crude telephones were sold in very small numbers to businesses and the exchange (central office) was technically very simple, with human operators plugging cables into and out of junction boxes. In due course the quality of reception improved and automatic exchanges were invented. Both these improvements required a great deal of development effort and many additional inventions and innovations. The telephone found it difficult at first to become widely accepted both because the quality of speech reproduction was poor but also, more importantly, because its usefulness was limited when the number of subscribers was small. In due course more telephone cables were laid and the number of subscribers increased. To lay cables and build exchanges was an obvious path for investors and innovators to take.

With global expansion of the telephone network the pace of development accelerated. Transmission became better, individual telephones were of better quality and more streamlined design, and it became possible even for long distance calls to be connected automatically. When everybody could dial everybody else throughout the world, the development of the stationary telephone had more or less reached its limit. However, soon the function of the telephone was supplemented by data transmission and the race was on for faster reliable transmission of increasing quantities of digital data. With this change the voice transmission tended to become digital as well, although the process of digitalization is not complete yet.

There was another possible way of expanding demand for telephones. Why have a telephone only in the office and in the home, when it became technically possible to have a telephone wherever you went. The mobile 'phone was brought onto a market that never knew it needed it, but accepted it with alacrity. The mobile telephone went from strength to strength and entrepreneurial firms soon added data transmission, access to the Internet and built-in digital cameras to mobile 'phone devices. At the moment industry is trying to sell TV reception on mobile 'phones, but whether the public will buy that is an open question.

Before the telecommunications industry could reach its present state, numerous technical inventions had to be made; from the humble magnet in a traditional telephone to sophisticated cell technology in the modern mobile 'phone. Thus the system was complemented and improved throughout its life, as well as expanded into performing additional tasks, not previously thought of.

The system nature of technologies provides valuable guidance to would-be innovators. The filling in of gaps in a system, the improvement of performance of certain parts of a system and of the system as a whole, and, finally, expansions of the system into previously uncharted territory all offer opportunities for innovators. The decisions what to buy and what not to buy are left to the market, i.e. to purchasing decisions by managers in the case of manufacturing technology and to purchasing decisions by individual consumers in case of consumer goods and services. Thus the market reigns supreme, as contemporary economic and political theory demand.

In reality, however, the market is distorted by all kinds of mechanisms. First and foremost, large firms create monopolies or oligopolies that dictate not only prices but also technological trends and the pace of technological change. It is obvious that industry marches more or less in unison. Major change in technological systems occurs more or less synchronously throughout an industry. Secondly, the actual market is not determined by genuine demand but by artificially created demands. A vast and sophisticated public relations and marketing apparatus invests enormous sums of money, manpower and ingenuity in telling people what they want to buy. To integrate a digital camera into a mobile telephone might seem like an exotic whim, but once the idea is pushed hard enough by the advertisers, it advances from whim to apparent necessity. A mobile 'phone without a camera suddenly becomes old-fashioned and obsolete. I suppose the combined device can be marginally useful to a few people who might take on-the-spot pictures of, say, real estate to send back to their offices. But does it offer any real utility to the youngster who would not be seen without it? The role of fashion in purchasing decisions cannot be overestimated. The me-too factor is as strong in the ownership of technological devices as in the ownership of anything else.

What happens when social needs have no champion? What happens about the urgent needs for technology that safeguards the natural environment? If governments are not prepared to look after the environment and provide the cash and the incentives needed to produce environmentally benign technologies, who else will? The task is too large for voluntary organisations and can only be very inadequately broken down into private purchasing or behavioural decisions. Governments are there to safeguard the public good and it should therefore be mandatory for governments to protect the environment. Many governments do little more than pay lip service to the task; and only some really try to do something substantial. Some governments of the greatest environmental polluters do not even pay lip service to the need for environmental protection and hide behind the convenient theory that the market knows best; the market will do all that is necessary. This is both logically and practically a nonsense and means that if humans want to bring about their own destruction, so be it. Humans do not really deliberately wish to destroy the planet but are unable to prevent it unless their major social institutions, such as governments, take over this task. National governments and international governmental institutions provide the only possibility of taking a technological path toward a sustainable future. It is often argued that the necessary measures are too expensive. They are not. All that is required is the development of environmentally benign technologies. Such technologies can provide ready markets for the right entrepreneurs. Governments merely need to prime the pump and, in some fields of technology, e.g. energy, provide the right regulations and, occasionally, finance major schemes such as tidal power stations. Waging war to acquire energy resources is no substitute for developing alternative energies and energy conservation schemes.

Currently the all electric automobile is being hailed as the saviour from carbon dioxide pollution and hence the answer to all worries about climate change. The argument is based on a very obvious fallacy. The proposed system consists of a battery and an electric motor. The battery will supply the electric current for the motor that will drive the car. The battery[18] will be charged from the normal electricity supply grid. But how will the electric power in the grid be produced? If it is produced from fossil fuel, then the carbon dioxide pollution is simply shifted from the internal combustion engine in the car to the power station feeding the grid. [19]I find it hard to believe that the proponents of the electric vehicle are unaware of this problem; it is too obvious. I suspect that the idea behind the all electric car is being pushed by the nuclear lobby because nuclear power generation, with all its risks and problems, does not cause carbon dioxide pollution. The hybrid electric car is a different proposition because here the battery would be charged by an internal combustion engine that also drives the car under certain circumstances. The alternative of using so-called fuel cells in which hydrogen is used to produce electricity for driving the vehicle is a promising proposition, perhaps the most promising, but still some way from becoming useful in practice. The problem of providing a network of hydrogen fuelling stations is rather formidable. The hydrogen could be produced in places with an abundance of solar energy and transported to where it is needed.

Technologists strive to produce new or improved technologies in order to re-kindle demand in often saturated markets. Two fundamental features of the interaction between humans and their technologies help to make never-ending innovation possible. The first feature is human greed in general, and the desire to buy more and more products of technology in particular. The second feature is an obsessive desire of humans to follow the latest fashionable trend, to keep up with the Joneses, to be up-to-date. Once a new technology gets a foothold on the market and establishes a new fashionable trend, there is no looking back. People want to have the latest and manufacturers make sure that the latest replaces the established as soon and as often as possible. Yesterday's latest cry of technology is today declared obsolete and somewhat pathetic compared to the very latest technology. New technology is, in a very real sense, addictive. Many people have a craving for the latest technological products. They regard it as shameful to use any mobile 'phone that is not up to the latest standard; they cannot be without the latest in music recording and playing technology; they must have the latest washing machine, cooker, or whatever. This addiction is, of course, aided and abetted by the advertising and marketing fraternity and ably supported by the media.

Each new technology does, of course, have some advantages over its predecessor. The advantage may, however, be quite small and inessential, yet clever marketing will kindle a general desire to own this latest product rather than stick with the older one. Some innovations are merely stylistic, but we should not underestimate the influence and power of fashion well beyond the field of clothing. Fashion dictates much design of consumer articles and, in particular, domestic styles in decorations, furniture, tableware, and even gardens. Fashion also dictates our taste in the exterior appearance of technological products such as motorcars, buildings, computers, TV sets and many more. More importantly, some small improvements in technology become fashionable and kindle the desire to purchase the novelties. Take, for example, the introduction of central locking or electric window winders in motorcars. Admittedly they offer small improvements in comfort, but their irresistibility

[18] The most promising candidate for a suitable battery is a lithium–ion battery, though this is not entirely unproblematic. If the use of such batteries becomes widespread, lithium will become a valuable commodity.

[19] Admittedly a power plant produces energy somewhat more efficiently than a small internal combustion engine, so a small advantage can be achieved by the electric car.

mainly shows the power of fashion. Soon after the introduction of these items it became impossible to sell new cars not equipped with these new goodies.

Some improvements within the system 'motorcar' are more important and make more serious contributions to the safety of cars. Items such as ABS (preventing the blocking of wheels) and electronic stability programmes probably make valuable contributions toward car safety. Air conditioning, especially in its automated form, certainly makes a valuable contribution to comfort but, alas, uses up some energy.

The nature of technology as an assembly of systems, and the nature of humans to want the most fashionable product, makes life a little easier for the innovator. The task is nevertheless difficult and risky. There is a long way from idea to prototype, from prototype to marketable product, and from there to large-scale production. The road is long, full of obstacles and unexpected problems, and requires a great deal of money and perseverance. The vital question of whether or not the new technology will be profitable remains unanswered till a very advanced stage of the innovation process.

We have said that technologies are addictive and this is true in more ways than one. Once somebody has owned a car, he or she does not want to be without one. The addiction is reinforced by social adaptation to the motorcar. Public transport is usually neglected because it struggles to achieve economic viability in the face of competition from motorcars. Thus relying on public transport is often inconvenient and this inconvenience increases the attraction of car ownership. The car is still addictively attractive, despite all its environmentally negative impacts, despite the horrible carnage it causes on the roads, and despite the frequent gridlock it causes on the roads. From the point of view of society, the motorcar is a disaster; yet the illusion of power and freedom that it provides to individuals makes it irresistible.

Other technologies have proved equally addictive. Anybody who has ever owned a domestic washing machine is most reluctant to make do without it. The washing machine provides independence and frees people from real drudgery. Anybody who has ever owned a television receiver is most reluctant to abandon it. The addiction to television is very powerful and of socially dubious value; though a detailed discussion goes well beyond the scope of this book. I mention it only as an example of the addictive power that technology can have.

The natural inclination of most humans to want the latest technologies is, of course, reinforced by the manufacturers and their marketing departments. The consumer was compelled to change from long-playing records to compact discs by the simple expedient of virtually stopping the production of the older records and record players once the compact disc had more or less established itself on the market. The same is happening with video tape recorders that have become virtually unobtainable because they are being replaced by DVD recorders. The change of analogue TV broadcasts to digital broadcasts takes a little longer because the necessary investments are very large and because TV services have the status of quasi public services. The change will, however, be forced on the public by the same simple device of discontinuing analogue services and replacing them by digital services within the next very few years. People will have the option of not having TV at all or to avail themselves of digital terrestrial services, digital satellite services, or digital cable services. So embedded in our society has television become that very few people will forego it altogether, the overwhelming majority will choose some combination of the new service possibilities and will buy the requisite hardware to receive the digital programmes.

The pressure to purchase the latest medical equipment is somewhat different. There is a constant stream of innovations, with existing medical equipment being improved for greater effectiveness, greater accuracy and greater user comfort. In addition to improvements to current equipment, new machinery is added all the time to the range of diagnostic, surgical and therapeutic devices. Some innovations may be of marginal utility and some may not be strictly useful at all, but on the whole, equipment is constantly improving and expanding the range of applications. Providers of medical services are obviously under great pressure to ensure that their equipment is up to the latest standard. Many parts of medical provision are competitive and must therefore be able to boast of the best available equipment. Other parts are under public control and purchases are controlled by the public purse, but even these are under political pressure to provide the best possible service. Hovering in the background are the lawyers trying to threaten medical practitioners with legal action if there is even the slightest whiff of negligence. Using antiquated equipment is a weakness that medics do not dare to show, thus playing into the hands of sales departments for medical equipment.

Many small technical gadgets, including personal computers and amateur cameras, have to be replaced frequently because they develop faults. Repairs are inordinately expensive compared to new goods because industrial production is extremely efficient, whereas repairs are labour intensive and cannot be streamlined in the same way as production. The consumer often purchases new equipment to replace faulty items because the cost of repairs is exorbitant compared to the cost of buying new. An additional problem is unavailability of spare parts for older equipment and the incompatibility of old equipment with new accessories, or vice versa. Incompatibilities of all kinds and rapid obsolescence affect mainly computers and other electronic equipment.

From the point of view of manufacturers, rapid obsolescence, expensive repairs, and unavailability of spare parts for older equipment are viewed as blessings. Manufacturers, particularly motor manufacturers, used to push obsolescence to absurd levels. Under public pressure they were forced to increase the longevity of their products and these have now reached fairly reasonable levels. Household machinery still has too short a lifespan and computers age at a ridiculously rapid rate. It is argued that computers are a relatively new technology that is still undergoing rapid improvements. This may be so, but, on the other hand, ordinary consumers hardly benefit from the sophistication and power of the most recent machines and would be happier if their older computers lasted longer.

The rapid development of medical technologies poses substantial problems of funding medical services and also extremely controversial and difficult ethical problems that do not lend themselves to rational analysis. The problems I am referring to may be put in a nutshell: should medical technology be used to prolong life, of whatever quality, at all cost and under all circumstances. In my view it is very doubtful whether hopelessly sick people suffering total incapacity should be kept in a state of some kind of animation by the use of elaborate machinery. These are very problematic issues and solutions can only be found on an individual basis. There can be no universal answer to the question of what kind of life is worth living and should therefore be sustained. Nevertheless, the question needs to be asked from case to case and cannot simply be rejected as an inadmissible one.

There is a second ethical question associated with high tech and high cost medicine:

If financial means are not available to offer every available treatment to every sufferer from severe disease – as is the case in publicly funded health care – how do we ration funds? What criteria should be used? The degree of suffering; the age of the patient; his or her family responsibilities; his or her social importance? In this connection we might also question the right of the rich to buy themselves every kind of medical provision, even if the provision itself is scarce. There may not be enough specialists or enough machines to treat everybody; do the wealthy in this situation have the right to preferential treatment?

A thorough discussion of these complex controversial issues is well beyond the scope of this book. The reason I raised the issues is to show how deeply technology affects our lives. Indeed even the most searching questions about the meaning and value of life, on the face of it entirely unrelated to technology, need to be discussed in conjunction with the use of contemporary technology.

So far we have discussed technology mainly from the point of view of the consumer. We now turn to the manufacturer. The criteria applying to purchases of process or manufacturing technology are entirely different from those applying to consumer technologies. The main pressure on the manufacturer is the pressure of competition. This forces the manufacturer to do essentially two things: a) to produce technological innovations to keep ahead of the competition and to reap extra profits on innovative technologies, and b) to keep the cost of production as low as possible.

Although technological innovation is expensive and risky, it is, at least in high tech industries, mandatory. Only by innovation can any manufacturer gain substantial advantages over the competition and, of equal importance, innovative technology in its early stages can often command premium prices on the market and thus help to repay the considerable costs incurred in producing the innovation.

As technologies mature, the pressure on prices intensifies because more often then not an overcapacity develops because of over-investment in a new promising technology, and because markets tend to saturate once the technology is mature. Price competition becomes severe under these circumstances. The manufacturer is therefore forced to reduce costs in order to maintain a profit margin. The first thing that can be done to reduce costs is to improve the production technology. This means investing in better machinery and equipment that can improve the quality of the output, reduce the number of rejects, and reduce the requirement for labour. The product can be designed for ease of manufacture, thus improving the overall efficiency of the process. Apart from improving the design of the product and improving the manufacturing technology, thus causing savings in materials and energy, the manufacturer must use possible savings in one of the greatest cost factors: labour. Labour can be saved by better production technology and also by better organisation of the whole process of production, including items such as layout of the factory, stock-keeping, streamlining the delivery of parts, choosing efficient suppliers for parts and services, and so forth. The unhappy outcome of all this drive for efficiency is a reduction in the workforce while maintaining the level of production and sales.

Another method of reducing costs is to reduce the wage-bill. In the most advanced countries this is difficult to do because of old-established rights of workers and strong trade unions. As unemployment grows, however, so the pressure on wages and privileges of workers increases and effective wages can be reduced. The threat of moving production to lower wage countries is another effective means of reducing domestic wages, but the actual move of production to low wage countries produces the ultimate savings in labour costs. It is a well-established practice of Western companies to shift production to Asian or east European countries where wages are low and the work force is well educated, well disciplined and docile. The individual firm may benefit from such a move, but the economy of wealthy countries suffers. Unemployment rises, purchasing power declines and the sense of well-being declines. People need work and this need is greater and more fundamental than the need for ever-cheaper goods. And not everybody can make their living from providing those ubiquitous financial services. In any case, recent events have shown the vulnerability and insanity of much that is going on in the world of finance.

The firms using all these various cost cutting methods may prosper and the consumer may benefit from reduced prices of goods. However all these measures cause reductions in employment in the wealthy countries and indeed unemployment has become endemic in most of them. This is a very unfortunate development, as the unemployed suffer not only considerable financial hardship but also severe social and psychological problems. In a very real sense the unemployed are outcasts and their potential as constructive and valuable citizens is lost to society, much as their potential for work and their potential as consumers are lost to the economy.

Work has become a commodity in short supply and work is the one thing people need more than various fanciful goods. However, governments find it very hard to ensure full employment. This is not surprising in economies that are run on purely capitalist principles in which governments have very few possibilities of intervening in the economy. They use various programmes ranging from training to underpinning some employment schemes, and they reduce workers rights, but the effect of all these schemes is marginal and cannot solve the root-cause of unemployment. The root-cause is that production efficiency is very high and that there is too much production capacity for any sustainable market demand. Innovations help to keep firms and the economy going, but most economic growth now is so-called jobless growth. This means that production can rise without employing additional labour. This is a consequence of highly automated production machinery and of increased administrative efficiency owing to the use of computers.

There are alternative means of reducing the pressure of competition and thus the pressure on employment in the rich economies. One alternative is the introduction of various trade restrictions, such as monopolies or cartels. Because of the current fervent belief of most governments in the virtues of free trade, these measures are now much more difficult to use, though very few people will claim that they play no role in contemporary economies. With so many huge global firms exerting their considerable power over markets, it is hard to believe that competition and free trade work to such enormous benefit to the consumer as our politicians would have us believe. Considering the effects of Western trade policies upon developing countries, the faith in the benevolence of free trade, as defined by Western governments and the World Trade Organisation, is even harder to sustain.

Unemployment has become endemic in developed countries and catastrophic in developing countries. Reluctantly, we must leave developing countries out of our discussion – their huge problems do not fit into the framework of this book. Most economists in the Western world believe that the endemic unemployment problem would disappear if our industries became even more efficient, if our labour costs, especially costs such as social, pension, and health insurance, could be reduced and if labour became more flexible (read less demanding) and better qualified. All these measures might, just might, have a positive effect upon the supply side of the economy, i.e. upon manufacturers and service providers. On the other hand, workers who are less well secured in case of unemployment, sickness and old age would lose some of their propensity to spend. If wages were lowered as well, it is hard to see how the more efficient supply side, with an even greater production of goods and services, could find consumers willing and able to buy all these additional outputs. Improving the supply, without a corresponding increase in demand, is futile.

There are demands that remain unfulfilled despite unemployment and spare capacities in many industries. The unfulfilled demands are of two kinds:

First, there is a great deal of unsatisfied demand for various labour intensive services, such as health services, public transport, education, care for the elderly, sick and disabled, and so forth. The problem with these services is that by their very nature they are expensive and it is hard to see how they could be delivered with increased efficiency. Those in need of the services generally cannot pay their full costs. Hence the public purse is required to face the bill and the public purse cannot, or will not, stretch to covering the full potential demand. Solving this problem would create massive employment and create a more humane society. At the moment, it is hard to see how this might happen in this profit-oriented society.

Secondly, there is considerable potential demand for environmentally benign and sustainable technologies. They range from very simple measures, such as more intelligent flushing systems for toilets to save water, to the highly complex and expensive projects for sustainable non-polluting energy supplies, to technically feasible but socially complex measures to reduce pollution from motorcars and lorries.

Energy supplies are a highly controversial topic. The positions range from that of the US, the greatest energy guzzler and polluter of them all, whose government simply does not want to know about measures to reduce greenhouse gases[20] or measures to curtail the use of nuclear energy, to that of the German Government that is committed to phasing out nuclear energy. France, on the other hand, is fully committed to nuclear energy, whereas the British Government has announced yet another review of energy options with the likely outcome that a new generation of nuclear power stations will be built. The nuclear option is advocated by many as the best option for large-scale electricity production because, unlike thermal power stations, it produces no greenhouse gases, in particular no carbon dioxide. Opponents of nuclear energy point out that it is expensive, that the future of Uranium supplies is uncertain, and, most importantly, that the radioactive waste produced by such power stations needs to be disposed of safely for a thousand years and that this is fraught with dangers and difficulties. They also point out that the power stations themselves might be prone to dangerous breakdowns, accidents and leaks of radioactive materials and might be vulnerable to terrorist attacks. Supporters of nuclear energy think that all the problems can be solved and that nuclear energy is the environmentally most benign and also the most economic solution to rising energy requirements.

Opponents of nuclear energy generally argue that there are plenty of non-nuclear possibilities to produce energy in environmentally benign and sustainable ways and also that our profligate use of energy could easily be curbed by suitable technological and social measures. Let us start with options for energy production. The order in which I list them is not necessarily their order in importance, as this is hard to gauge as yet. I start with the bio-fuel option, in which fuel is produced from agricultural crops of various kinds, beginning with fast growing woods and ending with rapeseed or sugar beet. Almost any plant and much organic waste can be converted into fuel by suitable fermentation processes. The fuel can be burned in conventional ways and the burning process does, of course, produce some carbon dioxide. However, as the plant uses dioxide from the air to grow, the net input of dioxide into the atmosphere is zero when burning fuel produced from plants. We may argue that bio-fuel is one way of using solar energy, because the plant uses solar energy to grow. The major objection to bio-fuel is that it uses agricultural facilities and soil that might otherwise be used for growing food, and thus drives food prices up beyond the reach of the poor. There can be no objection to the production of bio-fuel from waste. There are other ways of using solar energy, the ultimate source of most energy on earth. We can install solar panels of two kinds: either panels similar to central heating radiators that use the heat of the sun to heat water, mainly for domestic hot water systems; or we can use solar panels that convert the light

[20] This situation appears to be changing with a new President and a new administration having taken office recently.

from the sun directly into electric power. This, at the moment, is suitable for small-scale energy production, but is very useful for isolated places needing small amounts of electric power and is being technically improved all the time. We can also use solar energy much more indirectly, by using wind power or wave power. Both these technologies are used currently on a reasonably large scale. Some people object to the use of wind energy, claiming that it is unsightly, noisy and disturbs birds in their flight. Beauty lies in the eye of the beholder, the noise is not obtrusive and the degree of disturbance to birds is largely unknown. Another way of using wind power is to install devices that ride the ocean waves and translate their movement into rotary motion that drives an electric generator.

The next source of energy is not the sun but, rather indirectly, the moon. We can use tidal flow to produce energy either by simply inserting underwater turbines into the flowing tidal water, or, on a much larger scale, by building a barrage and thus capturing, in suitable estuaries, very large flows of water and, thus, produce large amounts of electric power in more or less conventional hydro-electric power stations. Some such schemes are in operation in Europe; in Britain a commission investigated the possibility of building a barrage across the Severn Estuary near Bristol. So far, the scheme has not been realised on account of the high cost involved and, less significantly, because of objections that local wildlife might be disturbed.

Finally, there is the hydrogen option. Hydrogen can be produced from natural gas, but that defeats the object of preserving the dwindling resources of gas. Hydrogen can also be produced from water; simply by splitting the two atoms of hydrogen from the atom of oxygen (water consists of H_2O). This can be done by electrolysis, which uses electric energy. The electrolysis needs to be powered by "clean electricity", that is electricity produced without carbon dioxide emissions. Hydrogen can be used as a mobile fuel with the attractive property that its sole product of combustion is water. The production, storage and transportation of hydrogen pose considerable problems. It appears likely that countries with plenty of sunshine will export this energy in one form or another; perhaps in the form of hydrogen or perhaps in the form of electric power.

Apart from the various possibilities of producing energy, we can also be less profligate in our use of energy. Houses can be better insulated; cars can be more efficient and need not travel so fast; public transport is more efficient in its use of energy than private personal transport, industrial production can be made more energy-efficient, and we need not use so much packaging. The fuel consumption of motor vehicles rises very sharply with increased speed of travel. Decreasing the speed of travel saves precious fuel and thus reduces costs and, of equal importance, it saves lives because it reduces the number and severity of accidents. We can save water and thus preserve a valuable resource, and thus also save energy used in pumping, cleaning and heating it.

Whether we speak of alternative sources of energy or of energy savings, in all cases there is scope for technological innovation with all its benefits of producing new employment. Recently there has been talk of collecting carbon dioxide as it escapes from chimneystacks and pumping it into some kind of storage facility. This would reduce the amount of carbon dioxide that enters the atmosphere, and may thereby reduce a further increase in the concentration of greenhouse gases, and thus avoid acceleration in climate change. Old oil wells or coalmines are among the storage possibilities being discussed. For some countries it may well be worthwhile to continue the production of coal and the operation of coal-fired power stations. They would need to make sure that the waste gases are thoroughly cleaned and consider the possibility of storing carbon dioxide, which is an unavoidable product of the burning of coal and other fossil fuels. The days of coal may not be over yet.

Whatever we might do about producing alternative energy, we certainly cannot get by without decreasing our energy consumption. There is not enough potential for the development of alternative energy sources to feed our profligate use of energy. We must save in order to make a real impact on the greenhouse effect and the potentially disastrous human-made fast climate change.

There would be considerable scope for work on environmental improvement and on redesign of technologies to make them environmentally more benign. Unfortunately the public purse is not very generous at the moment and environmental projects are starved of cash. To design products for environmental compatibility is not generally regarded as worthwhile, because the public is not inclined to spend money on such improvements unless they offer individual advantages in addition to the societal advantage of environmental compatibility. Perhaps entrepreneurs should be a little more inventive and offer environmentally benign technologies that do benefit the individual purchaser. Solar panels, for example, can reduce the energy bill of households. They are rather expensive, but in countries where government supported the technology in its infancy, thus bringing down the price, they find a ready market. House insulation is another technology that reduces household bills. Advanced central heating boilers and installations, with suitable electronic controls, are other examples. As the market for so-called hybrid cars shows, the public is not disinclined to purchase vehicles with very low fuel consumption if they become available at the right price and the right levels of comfort. Unhappily, at the same time as people buy frugal cars, they also buy ludicrous gas-guzzlers with engines that could propel a tank and acceleration and top speed suitable for racing cars rather than road vehicles. This development is, I think, one that calls for regulatory action that would remove such anti-social monsters from public roads. Automobile manufacturers will produce anything that provides them with profits and thus the frugal very small car lives side by side with extreme luxury cars or extremely fast so-called sports cars.

All kinds of socially desirable services are difficult to finance. At the moment discussions about the financing of retirement are en-vogue, whereas discussions on the financing of care for the elderly and the disabled do not feature quite so prominently in our media. Yet it is hard to finance such services because they are very labour intensive. With the demise of the extended family the elderly often end up in institutions. Neither the state nor private individuals find it easy to finance these institutions. They often run on a shoestring and provide inferior quality care, and yet they stretch private finance to the limit.

The situation of public transport is much the same. It seems almost impossible to run railway or bus services at a profit. The main reason for this is competition from private cars, but also very high costs of the necessary investment. So many problems of congestion in cities and of excessive pollution and excessive fuel-consumption would be eased if more transport of passengers in cities, and of goods between cities, could be shifted from motorcars and lorries to trains and buses.

Another socially desirable range of services that are difficult to finance are education in general and infant education in particular. The long fought for equality of women and equality of occupational opportunities for women, hinges on care for infants and toddlers. There is no known or conceivable technological "fix" for this problem despite a well-established social need. Not only is technology unable to take over from qualified people to look after children; it cannot even reduce the cost of such services. Except in horror visions, it requires humans to look after the human young. The cost of high quality care for young children is very high and this is a huge burden, especially for single parents. An important social problem that has no technological answer. The only feasible answers are social.

It is a shameful sign of our times that one of the pillars of civilization; the provision of a secure environment – has been badly eroded. We now need more security services than ever before and technology finds considerable scope in aiding the provision of security. Metal detectors are widely deployed, especially at airports, to detect firearms and explosive devices[21]. Security cameras are in widespread use to deter criminals and to help catching them. Biometric data augment the well-established use of fingerprints, and computers are able to check and verify such data at great speed. Biometric data can be used in place of simple passwords or simple keys. Many countries are planning to use such data in identity documents, including passports.

The disposal of waste has become a major headache and a variety of technologies are being developed to help solve this problem. One of the more modern approaches to waste management is recycling. This requires the sorting of waste into components that can be recycled and does not, of course, get rid of all the waste. The main items that can be recycled are metals, glass, paper, various plastic materials, and some textiles. Apart from this type of recycling we can also use garden and kitchen waste to produce compost and thus reduce the total amount of waste that needs to be disposed of and produce a valuable material for the garden that replaces some chemical fertilizers. The waste that remains after all that needs to be either buried or incinerated. In both cases it is helpful to remove toxic materials first and in more advanced waste disposal schemes items such as batteries, waste oil, paint and pharmaceuticals are disposed of separately. Incinerators can be used for district heating, but their design and safe operation are somewhat problematic. Burying waste can be useful as landfill, but suitable sites are difficult to find.

---

[21] The metal detector cannot, of course, detect explosives; it can only detect metal parts of a bomb. Sniffer dogs, on the other hand, can detect explosives and drugs.

There are many ingenious ways of dealing with problems of waste. Automobiles are now designed so that many recyclable or problematic components can be removed with relative ease, while the rests can be compressed for ease of storage and transport, and the steel can be recycled. The recycling of steel, glass and paper are old-established techniques, but currently more recycling techniques are being developed in response to awareness of the need to preserve materials and energy and also because landfill sites are scarce and the dangers of seepage of toxic materials are acute.

It is often claimed that the use of the computer has increased the efficiency of services by leaps and bounds. It is also claimed that the computer has enabled new types of services to be introduced. Both these claims imply that the computer has strengthened and expanded the service sector in the economy and has thus contributed to economic growth.

The above statements beg two questions. First, what does efficiency in the provision of services mean? It is easy enough to measure some kind of efficiency for services that are bought and sold in a market. In this case efficiency means the ratio of outputs to inputs, both measured in monetary terms. This kind of formal measurement has no meaning in terms of the quality of the service or its social desirability, it merely means that by putting in a certain amount of capital, labour, energy, and so forth a service of such and such a monetary value can be sold. A very large number, probably a majority, of services are not sold directly in a market but are either provided by public bodies or are so complex and multifarious that the overall efficiency or profitability cannot be ascribed to any particular aspect of the service in any meaningful manner. The efficiency of public services is very hard to measure, and what measures are used are largely arbitrary and subject to political influence. Politicians love to claim that they, unlike their opponents, can increase the efficiency of public services, but none of them can ever provide a precise answer on how they will do it or how they will meaningfully measure it. They all say they will cut waste, but what is wasteful lies, more often than not, in the eye of the beholder. Is it wasteful to provide a home help to old ladies and allow the help to chat a little with her lonely old client? Is it wasteful to plant annual flowers in borders of shopping centres or parking places? Is it wasteful to provide pedestrian zones in city centres? There is no end to possible questions, but a dearth of sensible answers. The truth is that the quantity and quality of public services are not defined and, hence, the efficiency of providing them cannot be defined either. Undoubtedly some rational measures can be taken to run a given service at lower cost in undiminished quality. But these sorts of measures are very limited and usually fully exhausted. Generally speaking, when costs need to be cut, either some services are cut or their quality is reduced.

This brings us to a discussion of quality of services. There are as many measures of quality as there are services; the general rule is horses for courses. Take, for example, a passenger transport service. What the customer wants to see is comfort, punctuality, frequency, reasonable costs, transparency of tariffs, ease of booking, ease of access, ready availability of information, assistance with luggage, and possibly some more. The computer can help with some of these aspects of quality, but the computer can also be misused to cut costs and diminish the quality of the service. The internet can provide useful information about timetables and tariffs of transport services. Alas, most websites are badly designed and difficult to use and hardly any can answer obvious questions such as: what is the best time to travel if I want the cheapest available fare? Not everybody has access to the internet and not everybody can handle websites. Telephone services will therefore remain indispensable for a long time to come and need to be adequately staffed. Some travel tickets are sold over the internet and the actual process of using the ticket to check in can be a nightmare. The choice between overstretched staff and long queues on the one hand, or difficult to understand and to follow procedures for self-check-in, is the choice between a rock and a hard place. How idyllic were the times when nice people sold you the ticket that was best for you and told you what to do next! Did these people really have had to be replaced by horrid machines? Even cashiers are being replaced by horrid self-service cash points. How idyllic were the times when one could speak to a shop manager or a sales assistant and obtain sensible answers to sensible questions, whereas now one speaks to somebody in a remote call centre who only knows as much as the computer he or she is facing. And that is, generally speaking, not a lot.

Some services, we might call them high-tech services, use highly complex machinery to provide the actual service and, in addition, use computers in their administration and their logistics. We consider two examples: a hospital and an airline. The technologies that are fundamental to these services are medical technologies on the one hand and airliners on the other. Both these technologies are immensely complex and their application requires many technical and organisational ancillaries and a lot of highly trained staff. In both cases the computer is ubiquitous throughout the basic technology and throughout the service.

Computer applications fall into two groups: the computer as an administrative/organisational tool and the computer as a technical component of some machinery. Examples are not hard to find. In the case of a hospital we can think of computer tomography that has a computer as a central component, and we can think of patient records as a use of the computer for administrative tasks. In the case of an airline we can think of numerous computer systems used to control the jet engines or the auto-pilots, and we can think of computers used for passenger reservations.

Almost all contemporary services use the computer for at least one category of application. The only exceptions are services organised on a very small scale, one man or woman businesses providing services such a gardening, domestic cleaning, hairdressing and the like. We tend to overlook and underestimate these services, yet they contribute a great deal to the quality of life of many people and provide a source of income to many others. In my view, we could do with many more of these types of services and could readily dispense with many glamorous "financial services" that provide excellent income to their practitioners and often poor value to their clients.

Whether through the influence of computers or not, we have become obsessed with measurement. We want numerical values for everything and then construct comparative tables, league tables, graphs, targets and percentages of the attainment of targets. Most of these figures, tables and graphs mean very little, but we cannot employ thousands of computers and thousands of graduates of business schools in public administration and in business and yet escape this particular penchant. Computers are good at it and business schools love it because it endows them with a kind of scientific respectability.

There are two fundamental difficulties with this kind of thing. Excessive measurement, often of items that cannot be meaningfully measured, and excessive setting of numerical targets, leads to distorted values and to futile goals. It leads to the pursuit of false gods and to misallocations of resources.

The second fundamental problem with the excessive worship of measuring rationality is that it is juxtaposed with irrationality. Whereas it is perfectly feasible to think and act rationally even in the absence of measured data; we now seem to think that the only alternative to measurement is irrationality. Even if such sentiments are rarely clearly enunciated, they are often applied in practice.

Unfettered competition and excessive reliance on advertising and the media has led to a loss of confidence. The modern citizen feels constantly cheated. Cheated by politicians, cheated by advertisers, cheated by the media, cheated by business and cheated by bankers.

We cannot blame the loss of confidence entirely on technology. There are many causes, but in many of them technology plays at least some role. The dominance of multi-national firms and very large business firms has caused considerable unease among the citizens and the faith in the integrity of such firms is at a very low ebb, to put it mildly. Technology plays a role in so far as without information and communication technologies such firms could not operate and could not have achieved their dominant positions. We may view technology as merely an enabling factor that allowed certain tendencies to become reality. But we may also think of technology as providing opportunities, ideas, and temptations on the line that anything that can be done shall be done, as long as it provides benefits to some skilful and well-placed players.

We have completely lost such faith as we ever had in the integrity of our politicians. We cannot blame technology for this, but must at least suspect that the media have something to do with it and the modern media are certainly creatures of technology. Politicians who are exposed to the daily gaze of the public on television find it hard, if not impossible, to retain any credibility. Too obvious are the twists and turns and implausibility of their arguments, too obvious their cavalier attitude to facts and the truth, too obvious their blatant attempts to ingratiate themselves to the public by saying anything that their public relations consultants think will go down well. Indeed it has in some cases become hard to distinguish between public relations and politics or policies and, as a result, the public feels constantly cheated.

There are more direct influences of technology that cause loss of confidence in citizens. First and foremost it is the loss of personal contact. We are no longer personally connected to large parts of our society; our connection is now via an intermediary and the intermediary is a machine – the computer. We can no longer telephone our suppliers directly; we have to go through a call centre and although we nominally speak to humans, in effect we speak to a computer. The human operator merely puts into human language what he or she reads off the computer screen. The feeling of the citizen is that he or she is dealing with a vast anonymous impenetrable organisation. It is the situation described by Kafka in his novel "The Castle" all those years ago. The difference is that Kafka described an impenetrable alien and undemocratic bureaucracy, whereas we are now dealing with supposedly friendly democratic institutions corrupted by the computer into something alien.

A blatant example of the generation of mistrust, with technology acting as a facilitator for mischievous and criminal human behaviour, is internet criminality. It ranges from the considerable nuisance, and often severe damage, caused by viruses and worms and whatever else these constructs of sick brains are called that infest the Internet, to downright theft perpetrated on the Internet, and includes criminal child pornography. The latter is made more dangerous by chat rooms and similar institutions that enable adult criminals to pose as youngsters who "innocently" wish to meet other youngsters. The Internet makes it easy to assume a false identity with criminal intent. The Internet thief is a very different animal from the common and garden burglar; the Internet criminal only needs good knowledge of computers and some ingenuity to manipulate other people's money for his or her own benefit. For protection against such criminals locks or burglar alarms or fierce guard dogs are of no use. What is required are ingenious security programmes that are constantly being improved. The old race between criminals and security devices is now being run between computer programmes. The police too has had to adapt to the new situation and needs computer experts to fight computer crime.

Another matter that causes uncertainty and ambiguity toward technology is the frequent malfunction and unreliability of many high-tech products. Manufacturers drive the technology to its limits and rush out new products before their development and testing is completed. They do this in an attempt to be first, or at least among the first, to bring out products with novel features onto the market because such products can be sold at premium prices and because high-tech firms lay great store by their image as being innovative. The image of pioneer is worth a great deal to them. The reverse side of the coin is the disappointment of many purchasers that their shiny and wonderful new equipment is nowhere near as wonderful as the marketing departments claim and, much worse, that equipment often fails entirely after a very short period, usually just after the expiry of the guarantee. And because manufacturing is very efficient, whereas repair work is labour intensive, repairs are inordinately expensive. All this adds to an atmosphere of frustration and mistrust; obviously things are not what they are supposed to be.

These comments apply to almost all high-tech sold to the ordinary consumer and the only beneficiary from the unreliability of products are various insurance companies who offer insurance against failure; albeit at a very high cost. The consumer is ambiguous; on the one hand he or she is attracted to the new apparently wonderful products, on the other hand their unreliability engenders a general feeling of being cheated. And this feeling goes well beyond high-tech products and permeates many aspects of the individual's relations with society.

Some technologies are socially more desirable than others. The practical application of this simple and uncontroversial statement poses two difficult problems. First, who decides and articulates the social desirability or undesirability of the consequences of a technology? Secondly, how are we to know beforehand what consequences will flow from the future application of a new technology? Technology Assessment attempts to answer the second question in order to inform the decision making process and, hopefully, achieve better decisions[22].

The answer to the first question depends on who sponsors the technology in question. Large-scale publicly funded or supported projects must be decided upon by due political process. What this process is depends on the particular governance of a particular country, but the decision on major public technological projects is, for better or for worse, an essentially political decision. Whatever we may think of a particular system of governance, the government of a country is the ultimate guardian of the public interest and must decide whether a particular technological project is or is not in the public interest. The actual decision-making process may involve a variety of mechanisms, including the setting up of a commission of enquiry, public debate, debate

[22] See e.g. Ernest Braun, (1998), Technology in Context, London, Routledge

among interested parties, parliamentary debate, specific research projects aimed at informing the debate, and so forth.

Technological projects sponsored by private firms are a matter for the firm to decide and the decision will be based mainly on an assessment of the earnings potential of the project. The firm will ask whether the project is within its capabilities, whether it will enhance the reputation and the growth potential of the firm and so forth, but the expectation of profit - the so-called bottom line - will be the decisive consideration. The firm will give some consideration to the social and environmental consequences of the proposed technology. It will do this in order to ensure that the proposed technology conforms to all applicable existing and foreseen regulations and standards. These considerations are mandatory; no firm can afford to ignore existing and legally enforceable regulations or standards. For a variety of reasons, the firm may add further social considerations. It may do so because its management is enlightened and committed to environmental protection or some other social ideal, or it may do so because it wishes to gain the goodwill of an enlightened public.

Many major projects undertaken by private firms are supported in various ways by governments, be it by direct subsidies, advance purchases, R&D in government laboratories, tax concessions, or other means. Whenever government is involved in a technological project, it ought to be up to government to perform the necessary technology assessment or, at least, make sure that the commercial partners in the project have carried out a satisfactory assessment.

The main question the manufacturer asks before introducing a technology, having made sure that it conforms to all standards and regulations, is whether it will sell and bring in a profit. An enlightened manufacturer often attempts to advertise his wares by showing that their performance is better than regulations require, that the product is indeed beneficial to the environment and is safer than safe. One of the ways to make products more desirable for environmentally aware consumers is to show that they consume little energy and that they can be recycled at the end of a long useful life.

In these neo-liberal days 'regulation' has become a dirty word. But sensible regulation is vital to the functioning of society and to the functioning of technological systems within society. The controversy should not be so much whether a degree of regulation is necessary, but what constitutes sensible regulation. To answer this question in individual cases is one aspect of technology assessment.

Regulation of technology has essentially three aims: 1. To ensure the safety of the users of the technology. This includes, for example, regulations about the safety of electrical appliances both in regard of preventing electrocution and of avoiding fire hazards. 2. To ensure that the environmental damage of the technology is limited to whatever standard is agreed by the lawmakers. This includes, for example, regulations about permissible levels of harmful emissions from motorcars or from factory chimneys. 3. To limit the danger and inconvenience a technology might cause to others. This includes traffic regulations and the allocation of wavelengths in the electromagnetic spectrum to different users of wireless communications. Regulations may change with new technologies and new scientific discoveries, but also in the light of public awareness and of political developments.

One form of regulation consists of the setting of standards for products by special institutions set up for this purpose. Standards have the dual role of ensuring the safety and functionality of certain technologies and, no less important, the interchangeability of products. If, for example, every manufacturer were to produce a different electric plug, the users would have no guarantee that the appliance they bought could be plugged into the sockets in their home. If a product complies with the standards applicable to it, the customer has an assurance that the product will perform its functions to a satisfactory degree and will not interfere unreasonably with other users or the general public.

We come to the second question asked above: how are we to know beforehand what consequences will flow from the future application of a technology? In the strictest sense, we must admit that we cannot know the future effects of a technology, indeed that we cannot know much about most aspects of the future. We need not be so strict, however, and may be able to achieve useful insights by trying to foresee the consequences in the fullest possible way, rather than simply trusting to luck or a highly partisan forecast by a committed proponent of a technology. The future is always uncertain, but by systematically trying to foresee the consequences of our actions we improve our chances to achieve our goals and to avoid various pitfalls. Humans are planners, they expend much effort on planning the future, and Technology Assessment is a method for improving the planning of technology.

An enlightened manufacturer attempts to foresee, during the early stages of R&D, as many effects of the new technology as possible in order to use the technology to best advantage and to be spared unpleasant surprises. Similarly, a public authority, embarking upon a technological project, seeks to foresee all the consequences that will flow from the implementation of the project. Much has been written about the methodology for TA, but beyond a very general universal methodology, it is necessary to devise the TA to suit the particular project in mind. "Horses for courses" is an appropriate rule for TA.

The general methodology helps to set out the problem in a systematic way. In summary, we may say that each technology assessment should consist of four basic steps: 1. deciding the scope of the investigation; 2. describing the technology involved, including rival and complementary technologies; 3. attempting to foresee the impacts of the technology or technologies, showing both positive and negative effects and identifying the affected parties; 4. looking at options for policies to be adopted with respect to the introduction of the technology and possible ways of promoting, directing or regulating it. This formulation is clearly more appropriate to public bodies than to commercial firms, but in essence the process of TA is the same for whoever plans to introduce a technology.

Several remarks need to be made. First, it needs to be emphasized that TA is an interdisciplinary activity. It requires technical expertise, as well as social, legal, economic and commercial knowledge. Thus the TA investigators must consult with as many experts and, of equal importance, with as many potentially affected parties, as appropriate for the particular assessment.

Secondly, it is difficult to know in advance what scope of assessment might be necessary. If, to take an historic example, we had been given the task of assessing the impact of the earliest semiconductor devices, the rectifier and the transistor, we might have come up with purely technical effects, such as compactness and low power consumption of future electronic devices; the need for retraining of electronic engineers, and possibly the need for developing new manufacturing techniques, but not much else. Nobody could possibly have foreseen the revolutionary developments in electronic devices, the huge expansion in computing and computers, the change from analogue to digital electronics, and all the vast gamut of changes in technology and in society that followed in the wake of the earliest steps in semiconductor electronics. This is truly an amazing story of the humble acorn growing into a mighty tree.[23] It requires a good deal of lateral thinking to know in advance what scope of TA should be attempted. And, most important, a single TA will often not suffice; the process needs to be repeated as the story unfolds, as new technologies emerge, as new knowledge accrues, and as the first impacts are beginning to be felt. Once a technology has gone too far, once it has become entrenched, it becomes very difficult to do anything about modifying its social effects[24]. We only need to think of the motorcar, or of the computer, or telecommunications, or the railways, or aviation, to see what is meant.

Finally, the activity of technology assessment is an advisory activity. Its aim is to inform decision makers, to warn them of risks and make them aware of opportunities. Whether in the public arena or in private firms and whatever the specific set-up, decisions on technology (as on any other matter) should be taken with the aid of the best possible information. Better informed decisions are likely to be better decisions in the sense that undesirable consequences may be avoided and positive opportunities may be taken[25]. On the other hand, we must acknowledge, albeit with regret, that despite all information decisions are often faulty and that no amount of objective information has the power to override irrational forces, prejudice and self interest. Even well established facts, let alone uncertain projections and controversial theories, are not necessarily the most important consideration in a decision. The power of science is limited in two senses. First, much knowledge that would

---

[23] For a discussion of the influence of cybernetics on R&D see Michael Nentwich, *Cyberscience*, 2003.

[24] David Collingridge, (1980), The Social Control of Technology, Open University Press, Milton Keynes, discusses thie problem in detail

[25] Nassim Nicholas Taleb argues in his book *The Black Swan* (2008), Penguin Books, that forecasting is useless because the really important outcomes are determined by unforeseeable events and circumstances

help decision makers is simply either unavailable or insufficiently well founded to be truly helpful. Secondly, inconvenient facts are often ignored by decision makers upholding their own interests or prejudices.

One factor that relegates the eminently sensible activity of Technology Assessment to a minor role is the globalisation of the economy and of technology. If a technological trend starts anywhere on earth, the rest of the world will follow it almost without question for fear of missing the boat and falling behind in the fiercely competitive quest for profit. Nevertheless, if the activity of TA is carried out with international cooperation it can still be very useful.

Because technological innovation is believed to stimulate the economy and to bestow competitive advantages upon countries, most governments try to stimulate technological innovation in a variety of ways. First and foremost, the public purse supports basic science, also known as curiosity driven science. This may be seen as a general cultural activity with the aim of knowing more about the world we live in; knowledge for the sake of knowledge, as a value in its own right. It may, however, also be seen as providing the fundamental knowledge on which future technologies may be based. We no longer believe that the distinctions between pure and applied knowledge are entirely clear-cut. Governments hope that their support for a purely cultural activity will, one day, yield economically useful knowledge. Governments do, of course, support applied research and technological development and, sometimes, even lend support during the early stages of the introduction and marketing of new technologies. The last of these may take the form of government purchases, or government sponsorships of conferences and seminars, or government help for pioneering early buyers of the new technology. Government also finances what we might call a research and information infrastructure, in the form of research institutes, research libraries, the patent system, learned societies, and a multi- layered education system. Government provides a variety of tax incentives for research and development and for technological innovation. And, last but not least, government itself – particularly its armed forces – is a major purchaser of innovative technology.

Before lending support to a new technology, government must make sure that the technology will bestow real benefits and will not cause harmful side effects. To do this, government or its agents usually perform a Technology Assessment or a series of assessments in various forms. Although it is not possible to foresee all effects of a technology in its infancy or, harder still, during its embryonic state, the attempt is nevertheless well worth the effort because it can also help to shape the technology to some extent in order to improve its utility and reduce its ill effects. What represents utility and what is an ill effect is, of course, debatable and must be subject to a political process. The TA can provide useful, careful and balanced information to feed into the decision process. Technology Assessments should inform both the regulation of technology and support for technological innovation by public authorities.

Currently, for example, there is a lively debate about the utility and the hazards of genetic engineering in all its aspects and ramifications, including genetically modified food. If we are to believe certain commentators, we might be at the beginning of a development that looks to me like a horror scenario[26]. Careful and numerous Technology Assessments will be needed before and during the process of lending public support to these developments and legislating in this highly controversial area. I hope that the Technology Assessments will stimulate and inform a public debate that will influence the political decision makers. Too much is at stake to leave the matter in the hands of scientists, ill qualified to make moral or political judgements. Too much is at stake to leave the matter to the lay public, uninformed in matters of science and technology. Too much is at stake to leave the matter to the discretion of politicians, driven by their own sets of interests. The future is too important to be left to the sole discretion of any of these groups and should be in the hands of scientists and engineers, technology assessors and social scientists, the general public and politicians, each contributing their shares and acting in cooperation.

Technology is inescapable; we cannot survive without it. We must make sure that it is used for the good of all humans and to help us take care of our planet.

26 See e.g. Pierre Baldi, (2001), The Shattered Self, MIT Press

Further Reading for this Chapter

Arthur, W. Brian. (2009). The Nature of Technology. London: Allen Lane.
Braun, Ernest. (1995). Futile Progress: technology's empty promise. London: Earthscan.
Braun, Ernest. (1998). Technology in Context: technology assessment for managers. London: Routledge.
Landes, David S. (1998). The Wealth and Poverty of Nations. London: Little, Brown & Co.
Pool, Robert. (1997). Beyond Engineering – how society shapes technology. New York: Oxford University Press.
Porter, A. L. and F. Rossini, R. A. Carpenter, G. Roper. (1980). A Guidebook for Technology Assessment and Impact Analysis. New York: North Holland.
Rosenberg, Nathan. (1994). Exploring the Black Box: technology, economics and history. Cambridge: Cambridge University Press.
Tichy, Gunther (ed.). (1996). Technikfolgen-Abschätzung in Österreich. Vienna: Austrian Academy of Sciences.
White, B. L. (1988). The Technology Assessment Process. New York: Quorum Books.
Winner, Langdon. (1977). Autonomous Technology: technics-out-of-control as a theme in political thought. Cambridge, Mass.: MIT Press.

Bibliography

Agricola, Georgius. (1912). De Re Metallica, translated from the first Latin edition of 1556 by Herbert Clark Hoover and Lou Henry Hoover. London: The Mining Magazine.
Angelucci, Enzo and Alberto Bellucci. (1975). The Automobile: from steam to gasoline. London: Macdonald & Jane's.
Arthur, W. Brian. (2009). The Nature of Technology. London: Allen Lane.
Ashworth, William. (second edition 1962). A Short History of the International Economy since 1850. London: Longmans.
Baldi, Pierre. (2001). The Shattered Self: the end of natural evolution. Cambridge, Mass: The MIT Press.
Baldwin, Neil. (2001). Edison: inventing the century. Chicago: University of Chicago Press.
Basalla, George. (1988). The Evolution of Technology. Cambridge: Cambridge University Press.
Brandon, Ruth. (2002). Automobile – how the car changed life. Basingstoke: Macmillan.
Braun, Ernest and Stuart Macdonald (second edition 1982). Revolution in Miniature – the history and impact of semiconductor electronics. Cambridge: Cambridge University Press.
Braun, Ernest. (1984). Wayward Technology. London: Frances Pinter Publishers.
Braun, Ernest. (1995). Futile Progress: technology's empty promise. London: Earthscan.
Braun, Ernest. (1998). Technology in Context:technology assessment for managers. London: Routledge.
Braun, Ernest and D. Collingridge and K. Hinton. (1979). Assessment of Technological Decisions-Case Studies. London: Butterworths.
Brentjes, Burchard and Siegfried Richter and Rolf Sonnemann (ed.). (1987). Geschichte der Technik. Köln: Aulis Verlag Deubner &Co KG.
Checkland, S. G. (1964). The Rise of Industrial Society in England 1815-1885. London: Longman.
Collingridge, David. (1980).The Social Control of Technology. Milton Keynes: Open University Press.
Coombs, R and P. Saviotti and V. Walsh. (1987). Economics and Technological Change. London: Macmillan.
Coulton, C. G. (1947). Medieval Panorama. Cambridge: The University Press.
Davis, Martin. (2001). Engines of Logic: Mathematicians and the Origin of the Computer. New York: W. W. Norton & Co.
Dosi, Giovanni, C. Freeman, R. Nelson, G. Silverberg and L. Soete, eds. (1988). Technical Change and Economic Theory. Pinter Publishers, London & New York.
Drucker, Peter F. (1970). Technology, Management, and Society. London: Heinemann.
Dyer, Christopher. (2000). Everyday Life in Medieval England. London: Hambledon and London.
Eden, Paul and Soph Moenk. (2002). Modern Military Aircraft Anatomy. Leicester: Silverdale Books.
Endres, Günter (ed.). (1998). Modern Commercial Aircraft. London: Salamander Books.
Engels, Friedrich, ed. David McLellan. (1993). The Condition of the Working Class in England. Oxford: Oxford University Press.
Fagan, Brian M. (fourth edition 1999). World Prehistory: a brief introduction. New York: Longman.
Fernández-Armesto, Felipe. (2000). Civilizations. Basingstoke: Macmillan.
Freeman, Christopher, John Clark and Luc Soete. (1982). Unemployment and Technical Innovation; a study of long waves and economic development. London: Frances Pinter.
Gamble, Clive. (1999). The Palaeolithic Societies of Europe. Cambridge: Cambridge University Press.
Gay, Peter. (1977). The Enlightenment: an interpretation. London: Norton Paperback.
Gordon, John Steele. (2002). A Thread Across the Ocean: the heroic story of the transatlantic cable. London: Simon & Schuster.
Halliday, Steven. (2001 new paperback ed.). Great Stink of London: Sir Joseph Bazalgette and the cleaning of the Victorian metropolis. Stroud: A. Sutton.
Hay, Denys. (1966). Europe in the Fourteenth and Fifteenth Centuries. London: Longmans.
Haywood, John. (1997). Ancient Civilizations of the Near East and Mediterranean. London: Cassell.
Heer, Friedrich. (1961, paperback 1998). The Medieval World- Europe 1100-1350. London: George Weidenfeld & Nicolson, paperback: Phoenix.
Hobsbawm, E. J. (1973). The Age of Revolution – Europe 1789 – 1848. London: Sphere Books.
Hodges, Henry. (fourth impression of 1989 edition, 2000). Artifacts: an introduction to early materials and technology. London: Duckworth.
Hodges, Henry. (1970). Technology in the Ancient World. London: Allen Lane The Penguin Press.
Holmes, George (ed.). (1988). The Oxford Illustrated History of Medieval Europe. Oxford: Oxford University Press.

Hornblower, Simon and Anthony Spawforth (editors). (1998). The Oxford Companion to Classical Civilization. Oxford: Oxford University Press.

Hughes, Edward (1995, 7th edition, revised by I McKenzie Smith). Electrical Technology. Harlow: Longman.

Jenkins, Lawrence. (1987). Digital Computer Principles. New York: John Wiley & Sons.

Jones, Howard. (1973). Steam Engines: an international history. London: Ernest Benn.

Klein, Naomi. (2007). The Shock Doctrine: The Rise of Disaster Capitalism. Allen Lane. Penguin Group, London-New York.

Klemm, Friedrich. (1979). Zur Kulturgeschichte der Technik. München: Deutsches Museum.

Klemm, Friedrich. (fourth edition, 1999). Geschichte der Technik: der Mensch und seine Erfindungen im Bereich des Abendlandes. Leipzig: B. G. Teubner Stuttgart.

Koestler, Arthur. (1959). The Sleepwalkers. London: Hutchinson.

Kuhn, Thomas, S. ((1957). The Copernican Revolution. Cambridge, Mass.: Harvard University Press.

Landes, David S. (1969). The Unbound Prometheus: technological change and industrial development in western Europe from 1750 to the present. Cambridge: Cambridge University Press.

Landes, David S. (1998). The Wealth and Poverty of Nations. London: Little, Brown & Co.

Lane, P. (1978). The Industrial Revolution. London: Weidenfeld & Nicolson.

Macdonald, Stuart. (1998). Information for Innovation: managing change from an information perspective. Oxford: Oxford University Press.

Mantoux, P. (1961). The Industrial Revolution in the Eighteenth Century. London: Jonathan Cape.

Mathias, Peter. (1969). The First Industrial Nation: an economic history of Britain 1700-1914. London: Methuen & Co.

Mazlish, Bruce. (1993). The Fourth Discontinuity: the co-evolution of humans and machines. New Haven: Yale University Press.

Meadows, D. H. and D. L. Meadows and J. Randers. (1992). Beyond the Limits: global collapse or a sustainable future. London: Earthscan.

Mokyr, Joel. (2002). The Gifts of Athena: historical origins of the knowledge economy. Princeton & Oxford: Princeton University Press.

Morazé, Charles (ed.). (1976). History of Mankind: the nineteenth century 1775-1905. London: George Allen & Unwin.

Musson, A. E. (1972). Science, Technology and Economic Growth in the Eighteenth Century. London: Methuen & Co.

Nader, Ralph. (1965) Unsafe at any Speed. New York:Grossman Publishers Inc.

Nentwich, Michael. (2003). Cyberscience: research in the age of the Internet. Vienna: Austrian Academy of Sciences Press.

Oakeshott, R. Ewart. (1960). The Archaeology of Weapons: arms and armour from prehistory to the Age of Chivalry. London: Lutterworth Press.

Pareti, Luigi with Paolo Brezzi and Luciano Petech. (1965). History of Mankind, vol. II, The Ancient World. London: George Allen and Unwin.

Pirenne, Henri. (sixth impression 1958). Economic and Social History of Medieval Europe. London: Routledge & Kegan Paul Ltd.

Pliny, the Elder.(1938). Natural History. Translated by H. Rackham. 10 volumes. London: William Heinemann.

Pool, Robert. (1997). Beyond Engineering – how society shapes technology. New York: Oxford University Press.

Porter, A. L. and F. Rossini, R. A. Carpenter, G. Roper. (1980). A Guidebook for Technology Assessment and Impact Analysis. New York: North Holland.

Porter, Roy. (second edition 2001). The Enlightenment. Basingstoke: Palgrave.

Postan, M. M. (1972). The Medieval Economy and Society, London: Weidenfeld and Nicolson.

Postrel, Virginia. (1998). The Future and Its Enemies – the growing conflict over creativity, enterprise, and progress. New York: The Free Press.

Pucher, John and Christian Lefèvre. (1996). The Urban Transport Crisis in Europe and North America. Basingstoke: Macmillan Press.

Ramelli, Augustini (1620). Schatzkammer Mechanischer Kuenste. Translated into German and published by Henning Gross, Leipzig. See also Ramelli, Agostino. (1976). Various and Ingenious Machines. English translation by Martha Teach Gnudi, Baltimore: Johns Hopkins University Press.

Rawcliffe, Carole. (1995). Medicine and Society in Later Medieval England, Stroud: Sutton Publishing.

Reinalter, Helmut (ed.). (1993). Aufklärungsgesellschaften. Frankfurt a.M.: Peter Lang.

Rosenberg, Nathan. (1994). Exploring the Black Box: technology, economics, and history. Cambridge: Cambridge University Press.

Rutkow, Ira M. (1993). Surgery – an illustrated history. St. Louis: Mosby Yearbook Inc.

Rybczynski, Witold. (2000). One Good Turn – a natural history of the screwdriver and the screw. New York: Simon and Schuster.

Scarre, Chris. (1998). Exploring Prehistoric Europe. Oxford: Oxford University Press.

Shirley, Elizabeth. (2001). Building a Roman Legionary Fortress. Stroud: Tempus Publishing Ltd.

Singer, Charles and E. J. Holmyard and A. R. Hall (ed.). (1954). A History of Technology, vol. I, From Early Times to Fall of Ancient Empires. Oxford: Clarendon Press.

Singer, Charles and E. J. Holmyard and A. R. Hall and Trevor I. Williams (ed.). (1956). A History of Technology, vol. II, The Mediterranean Civilizations and the Middle Ages. Oxford: Clarendon Press.

Singer, Charles and E. J. Holmyard and A. R. Hall and Trevor I. Williams (ed.). (1956). A History of Technology, vol. III, From the Renaissance to the Industrial Revolution. Oxford: Clarendon Press.
Singer, Charles and E. J. Holmyard and A. R. Hall and Trevor I. Williams (ed.). (1958). A History of Technology, vol. IV, The Industrial Revolution, c. 1750 to c. 1850. Oxford: Clarendon Press.
Singer, Charles and E. J. Holmyard and A. R. Hall and Trevor I. Williams (ed.). (1956). A History of Technology, vol. V, The Late Nineteenth Century. Oxford: Clarendon Press.
Volumes 6 and 7, the twentieth century, edited by T. Williams. See under Williams.
Sobel, Dava. (1999). Galileo's Daughter. London: Fourth Estate.
Street, Arthur and William Alexander. (10th edition 1994). Metals in the Service of Man. London: Penguin Books.
Tann, Jennifer (ed.). (1981). The Selected Papers of Boulton and Watt. London: Diploma Press.
Tichy, Gunther (ed.). (1996). Technikfolgen-Abschätzung in Österreich. Vienna: Austrian Academy of Sciences.
The Times History of War. (2000). London: Harper Collins.
Townshend, Charles (ed.). (1997). The Oxford Illustrated History of Modern War. Oxford: Oxford University Press.
Tunzelmann, von G. N. (1978). Steam Power and British Industrialisation to 1860. Oxford: Oxford University Press.
Uglow, Jenny. (2002). The Lunar Men – the friends who made the future. London: Faber and Faber.
Wangensteen, Owen H. and Sarah D. Wangensteen. (1978). The Rise of Surgery – from empiric craft to scientific discipline. Folkestone: Dawson.
White, B. L. (1988). The Technology Assessment Process. New York: Quorum Books.
White, K.D. (1984). Greek and Roman Technology. London: Thames and Hudson.
Whittle, Alasdair. (1996). Europe in the Neolithic. Cambridge: Cambridge University Press.
Wilkinson, Philip. (2000). What the Romans Did for Us. London: Boxtree.
Williams, Trevor I. (ed.) (1978). A History of Technology. Volumes 6 and 7, the twentieth century. Oxford: Clarendon Press.
Winner, Langdon. (1977). Autonomous Technology: technics-out-of-control as a theme in political thought. Cambridge, Mass.: MIT Press.
Winston, Brian: (1998). Media Technology and Society. London: Routledge.
Womack, J. P. and D. T. Jones and D. Roos. (1990). The Machine That Changed The World. New York: Rawson Associates.

Register of Names

Index

Curriculum Vitae

Ernest Braun, born in 1925, studied Physics at Charles' University, Prague (M.Sc.and Dr. Rer. Nat). Obtained Ph.D. in Physics from Bristol University, England. Worked as research physicist in industry and as lecturer in several universities. Became Professor of Physics at Aston University, Birmingham, England in 1967. Initially continued work in solid state physics, but became increasingly interested in questions related to the social significance of science and technology. Founded a post-graduate teaching and research unit, the Technology Policy Unit at Aston University and published many articles and books on the history of solid state physics, the mechanisms of technological innovation and the role of technological change in society. Retired from Aston University in 1984 and became head of the Technology Assessment Unit at the Austrian Academy of Sciences in Vienna. Retired in 1991 and spent a few years as visiting professor at the Open University, Milton Keynes, England. The present book is a kind of summary of his life's work.